Backyard Homesteading & Mini Farming

A Beginner's Guide to Growing Crops, Raising Chickens, Raising Goats, Beekeeping and Building Your Own Vegetable Garden for a Sustainable Living

By Luke Smith

© **Copyright 2020 - All rights reserved.**

The content contained within this book may not be reproduced, duplicated or transmitted without direct written permission from the author or the publisher.

Under no circumstances will any blame or legal responsibility be held against the publisher, or author, for any damages, reparation, or monetary loss due to the information contained within this book. Either directly or indirectly.

Legal Notice:

This book is copyright protected. This book is only for personal use. You cannot amend, distribute, sell, use, quote or paraphrase any part, or the content within this book, without the consent of the author or publisher.

Disclaimer Notice:

Please note the information contained within this document is for educational and entertainment purposes only. All effort has been executed to present accurate, up to date, and reliable, complete information. No warranties of any kind are declared or implied. Readers acknowledge that the author is not engaging in the rendering of legal, financial, medical or professional advice. The content within this book has been derived from various sources. Please consult a licensed professional before attempting any techniques outlined in this book.

By reading this document, the reader agrees that under no circumstances is the author responsible for any losses, direct or indirect, which are incurred as a result of the use of information contained within this document, including, but not limited to, — errors, omissions, or inaccuracies.

Table of Contents

BACKYARD HOMESTEAD 1

Introduction 2

Chapter One - Why Start a Backyard Homestead? 7
- *Having a Self-Sufficient Lifestyle* 9
- *Growing Vegetables, Fruits and Herbs* 14
- *Raising Farm Animals* 16

Chapter Two - Starting Your Home Garden 23
- *Preparing a Home Garden* 24
- *Raised Garden Beds* 28
- *Growing Vegetables, Fruits and Herbs* 31
- *What Plants Beginners Should Grow* 34
- *Herbs for Beginners* 34
- *Vegetables for Beginners* 36
- *Fruits for Beginners* 40

Chapter Three - Growing Grains on Your Homestead 47
- *How to Grow Grains* 50
- *Grains for Beginners* 57
- *How to Harvest, Process and Store Grains* 59

Chapter Four - Setting Up Your Barn 66
- *Building Your Own Barn* 68
- *Converting a Prefabricated Shed into a Barn* 75

Chapter Five - Raising Chickens 83
- *The Benefits of Raising Chickens* 84

Chapter Six - Raising Goats 97
- *The Benefits of Raising Goats* 99
- *How to Raise Goats* 109

Chapter Seven - Raising Cows 114

The Benefits of Raising Cattle *115*
How to Raise Cattle *117*

Chapter Eight - Raising Honey Bees 124

The Benefits of Raising Honey Bees *125*
How to Raise Honey Bees *131*

Final Words 140

MINI FARMING 143

Introduction 144

Chapter One - Why Mini-Farming? 148

Mini-Farming Requires Less Capital *149*
Mini-Farming Requires Less Space *151*
Mini-Farming is More Productive *154*
Mini-Farming is More Efficient *156*
Mini-Farming Gives You More Control *160*
Mini-Farming is Simple *162*

Chapter Two - Crop Farming 167

Monocropping *168*
Crop Rotation *172*
Mixed Cropping *176*
Intercropping *181*
Raised Garden Beds *188*
Hydroponics *196*
What Approach is Right For Me? *204*

Chapter Three - Profitable Vegetables to Grow 211

Growing Potatoes *212*
Growing Tomatoes *216*
Growing Cabbage *219*
Growing Onions *223*
Growing Herbs *225*
Jams and Salsas *226*

Chapter Four - Raising Specialty Livestock for Profit....... 232

Raising Cattle *233*
Raising Chickens *237*
Raising Goats *241*
Raising Bees *245*

Chapter Five - How to Profit from Mini-Farming............. 255

Profiting From Chickens *256*
Profiting From Cattle *258*
Profiting From Goats *260*
Profiting From Bees *262*
Profiting From Crops *264*

Chapter Six - Pest Prevention and Maintenance Control.. 270

Maintaining Your Crops *271*
Maintaining Your Livestock *273*

Final Words... 278

BACKYARD HOMESTEAD

A Practical Guide to Building Your Own Mini Farm & Raising Farm Animals for Beginners

By Luke Smith

INTRODUCTION

These days the cost of food continues to grow and people have begun to jokingly refer to the apocalypse. While these two events are unrelated, they share a similar DNA. No biblical apocalypse is about to happen but the world is changing and one of the things that have become clear is that there must be other ways to provide food for our families. We can't just rely on big business to farm and raise livestock for us.

Big business relies on chemical fertilizers to increase crop yield but this comes at a cost. These chemicals are introduced into our food and therefore our diets. We're not supposed to be consuming these chemicals… yet we do. And livestock is often even worse, with famous scandals involving growth steroids and maltreatment. The world has a growing sense that there are not only health consequences to these practices but also moral consequences that we are guilty of by our continued support.

For a time we looked past these issues. It was these very same practices that allowed us to eat a full meal at a much lower cost than the organically grown fruits and vegetables sold at farmers' markets. Meat was inexpensive due to these practices, too, and so a full plate (or a full stomach, rather) made up for any health or moral qualms we may have had.

But as prices continue to rise, we have to stop and ask ourselves if it is still worth the risk and the cruelty to animals that come with the territory. What's more, if you really do believe that the apocalypse is coming then what does that mean for our supply chains going forward? In all of our media and projections of that future, have we ever seen a still-functioning big business farm? No, of course not. If the world collapses, so too does industry.

While I personally don't believe that the apocalypse is anywhere near, the fear of such has actually had a profoundly positive effect on the world. Specifically, people have begun to learn to and practice growing their own vegetables and raising their own livestock. This makes me extremely happy to see because it means that people are learning to step away from predatory business practices to look after themselves and their families with their own two hands.

In this book, I am going to teach you how to get started making your own backyard homestead so that you can provide food for your family no matter what happens to the world around you. This way, you can always rest easy knowing that you have food. These skills are important for us adults to learn but I want to also implore you to teach these skills to your children. As parents, we want nothing more than to provide our children with what they need to stay healthy and live happily. If you teach these skills to your children, you teach them how to never go hungry. In my humble opinion, it is the greatest gift a parent can provide.

Each chapter will narrow its focus to look at how each piece of a backyard homestead is managed. When it comes to creating your own, remember that you don't need to follow each of these pieces of advice. For example, you might want to raise goats and chickens but choose to pass on cattle. But despite this, I do want to suggest that you read the chapter on cattle. While you might not follow it, knowledge is power and it will give you a grounding of understanding should you change your mind later.

Chapter one will introduce you to this concept in more depth. While it is easy to say that a backyard homestead will provide your family with food, what does this mean exactly? How do we achieve this goal? How much work do we need to put in? How much does it cost? How much maintenance does it require? Many of these questions will be answered more specifically throughout the book, as the cost to start a home garden or raise cattle are quite different. But chapter one will allow us to start to explore them and get a grounding on the subject that will carry us forward.

In chapter two we will see what it takes to start a home garden to grow vegetables, fruits and herbs. We'll look at how we can reduce the amount of space it takes to grow them, too, so that we can maximize our available space for larger yields or multiple crops. This focus on growing our own food will be followed up on in chapter three where we'll look at growing grains like wheat, rye or oats. Rather than simply give you a guide on how to grow one or two different edible crops, these chapters will focus on giving you the knowledge you need to grow anything you want. Yes, we will look at a couple in specific but this wider picture will ensure that you are well-grounded in the concepts of crop farming.

We'll move into what could loosely be considered the second part of the book as we enter chapter four. This chapter will look at how you can set up a barn of your own. Rather than look at building a barn, which would require us to learn carpentry, we'll look at how we can convert a shed into a barn. Remember that many areas don't allow you to erect a structure on your property without permission. However, if you already have a shed then you will be able to convert it without needing any paperwork. But just because you can convert a shed doesn't mean that it is legal for you to raise livestock on your property. Always check the laws in your local area, as well as in your state or province.

Chapter five explores raising chickens; chapter six looks at raising goats; chapter seven takes on the task of raising cows; chapter eight closes out the book with a look at how to raise honey bees. Each of these will look at the benefits of raising these animals, as well as the associated costs (both monetary and time-wise). I don't personally elevate any one of these animals above the rest, picking which is right for you is simply a matter of taste and logistics.

Together, these eight chapters teach you much of what you need to know about backyard homesteading. But there are still other options available to us for how we might go about this. As we only have so much space here, the final words will be used to suggest other ways in which you may decide to expand your backyard homestead in the future. With a little creativity and forethought, you'll find that there are many ways in which you can maximize your space, earn money from the practice, expand your diet and work in other productive angles such as raising sheep for wool or using aquaponics to raise fish at the same time you grow crops. So while the book in your hands is expansive, you should take it as the first step of many on your backyard homestead adventure.

CHAPTER ONE

WHY START A BACKYARD HOMESTEAD?

We've answered this question in a brief form already. The reason to start a backyard homestead is so that we can take control of our lives, specifically where our food comes from and how healthy it is for us to consume it. With hundreds of scandals coming to light within the agricultural industry, it should come as no surprise that more and more people have decided to take the leap into growing and raising their own food. It is a smart choice that they have made.

And by purchasing this book, you have demonstrated that you are also prepared to start making this transition. I won't lie and tell you that it will be easy. The truth is that it will take a lot of work. But the biggest piece of work comes at the beginning when we need to set everything up for the first time. Yes, there is lots of maintenance to be done to keep your backyard homestead functioning properly but many people, myself included, find this experience to be quite rewarding and relaxing. It might be hard work but it's hard work that leaves you with a sense of purpose, of belonging. There is just something about watching a crop grow to harvest or collecting fresh chicken eggs for your breakfast that fills you with pride. The fact that your two hands could achieve everything you need to fill your family's stomachs is just so powerful. You might not know what I mean yet, but you will shortly once you start to get your hands dirty.

In this chapter, we will start by looking at what it means to be self-sufficient and how a backyard homestead can help us get there. Being self-sufficient is about more than just growing or raising your own food, though this is a strong part of it. We'll follow this discussion with looks at why growing vegetables and raising livestock is so rewarding. To minimize the amount of repetition in the book, these sections will be brief. Consider them preparation for the chapters that look at them in more depth.

Having a Self-Sufficient Lifestyle

Before we can define what a self-sufficient lifestyle is, we should first define the concept of self-sufficiency. What does it mean to be self-sufficient? How can a family be self-sufficient when there are multiple people?

To begin with, we need to expand our definition of self. Rather than looking at ourselves as an individual, let us look at our family as a self-contained unit. We live together and look after each other. Within our home, we represent our own nation-state with rules set by those in charge and followed by the younger members. However, most families are not self-sufficient. They rely on food and income from other sources. Each member of the family can help to support one another, creating a higher level of self-sufficiency than someone who lives alone. For example, a college student can ask their parents for advice or help when they live at home but they need to reach outside of their home for that same help when they live by themselves. So for the sake of understanding, self-sufficiency as we'll define it sees the self as the household in question rather than the individual.

With that solved, what then is self-sufficiency? To be self-sufficient can be defined in two ways. One meaning relates to the confidence a person has. When someone is confident in their own abilities, they are self-sufficient. This definition might seem irrelevant for our purposes but we'll be touching on it again in a moment. The other definition is more in line with what you are probably thinking. That definition sees self-sufficiency as meaning that a person is able to care for all of their needs (food, shelter, etc.) without help from the outside world.

Without help from the outside world means that a person can make enough money on their own to have ends meet. They can cover their rent or mortgage, they can keep food on the table and they can pay the bills necessary to keep the lights and heat on. We often see these are the end-all signs of self-sufficiency but they shouldn't be. You could go to work at Walmart and make enough money to achieve these goals but are you really self-sufficient? Your food is still coming from the store, your money is entirely reliant upon working to earn a profit for a corporation that doesn't care about you and a recession could easily see you without a job. If you could lose your job at any moment, how sufficient are you? We have convinced ourselves that this is a form of self-sufficiency but we are horribly mistaken.

To live a truly self-sufficient lifestyle, we must take control of these factors. This is where backyard homesteading comes in. As you learn to grow crops and raise livestock, you are learning how to become more responsible for your own needs without the need for outside assistance. This is most easily understood when we consider food as part of self-sufficiency. Crops provide fruits, vegetables and herbs. Livestock provides meat, eggs and other products which can be made from the raw resources they produce. So backyard homesteading helps us to take control of our diets.

But what about that other need that was mentioned? Money.

to keep our home and our backyard homestead, we need to have enough money for rent and power. Surely this points towards a limitation of the backyard homesteading lifestyle, right?

Wrong.

There is a lot of money to be made this way. Crops can be sold locally, both directly to customers or to stores and restaurants. The products produced by your livestock can be sold but so too can the livestock themselves. In fact, if you decide to raise bees then you might be shocked to learn that you can even rent your bees out to local farmers. We'll get more into that in chapter eight, don't worry. But what this means is that this isn't just a way to provide food for your family. This can be a way to earn your income. Depending on how seriously you want to take backyard homesteading, it can easily become your main source of income. Now, it will require time and money to get going. So don't quit your day job just yet. But if quitting is a goal of yours then backyard homesteading offers one way through which you can make that dream a reality.

Now, remember how I said we would come back to that earlier point about confidence? Self-sufficiency can be defined as having total confidence in your abilities or your capability. I say capability because this one is a little broader. For example, you might not have the ability to provide medical treatment to your livestock but you can still be fully capable of seeking out and getting this treatment. This might sound like it blurs the line of self-sufficiency but it doesn't. Being self-sufficient doesn't mean you don't have to rely on others at times. We all need to get medical care or legal advice, we all have to rely on delivery men to bring us the tools and resources we've purchased. Being self-sufficient does not mean being cut off from the world or cutting ourselves off from making use of services that benefit us.

But knowing that we are able to use our own two hands to provide food and make a living instills a level of confidence that has to be felt to be understood. This confidence is further boosted by the knowledge that we fill our brains with. Following up on that last example, having the knowledge to recognize that your livestock are sick is a sign of self-sufficiency. Being able to recognize a problem and then reach out and hire the appropriate services to treat the problem is a sign of self-sufficiency. This point is extremely important because many people seem to think that self-sufficiency means entirely relying on the self for everything that they encounter. But this is not self-sufficiency, this is self-delusion. We aren't infallible after all.

Self-sufficiency, as we defined, means that we can supply our own needs. This means food, money, shelter. We don't need a vet to treat our livestock. If they get sick and die, we continue to live. We might lose money and have to tighten our belts but we don't lose home or health. It is these needs that the self-sufficient lifestyle looks to provide and it is these needs that we will be mastering throughout this book.

Growing Vegetables, Fruits and Herbs

Growing vegetables, fruits and herbs are the cheapest part of a backyard homestead. If you absolutely had to you could grow them with nothing more than a few seeds and the ground in your backyard. Of course, it is going to be a little bit more complicated than this if you want to do it properly but we'll be tackling that in the next chapter.

For now, let's consider why vegetables, fruits and herbs make for sure a great part of a backyard homestead. Of course, the cheap cost to get started growing crops is one plus that cannot be ignored. Many backyard homesteads first began as a backyard vegetable garden. But if this was the only benefit that growing crops offered then this would be a much shorter chapter.

The other factor that we can't look past is the simple fact that these crops provide us with nutritious and delicious foods for our dinner table. We all know that vegetables are high in nutrients and they make for a part of a healthy diet but what about fruit and herbs? Fruits offer us a way of indulging our sweet tooth. The natural sugars in fruit make them sweet and mouthwateringly tasty but without the need for processed sugars. We can get those same sugars directly from the source, essentially. And while herbs are used to spice up and add flavor to a dish, did you know that they are jam-packed with so many nutrients that they make vegetables look like slackers? On a pound for pound basis, the nutritional values of garden herbs are astronomically higher than vegetables. Of course, they shouldn't just be eaten raw on their own but adding them to your garden and thus your diet is one of the smartest choices you can make for your health.

But the health benefits and the cheap cost are only the two most obvious benefits. So what about the rest? Let us remember that we are talking about growing vegetables, fruits and herbs as part of a backyard homestead and not merely just starting a vegetable garden. When we raise animals as part of our homestead, we find that we have to deal with animal poop. On its own, this isn't very fun. But if you are also growing a garden then this means that you have direct access to free manure. Plus you can turn around and use a portion of the vegetables you grow to feed your animals, making it a circular relationship.

Vegetables are also a profitable venture. People have to eat and many like eating healthy, organically grown fruits and vegetables that haven't been polluted with chemical fertilizers. When you factor in the low cost to get started growing these crops, it is easy to quickly turn a profit with them. Being able to quickly turn a profit is important for those looking to use their backyard homestead as a way of paying the bills.

Many think of vegetable farming as a purely spring to fall activity. However, with the use of greenhouses or hydroponic approaches, you can grow pretty much anything anytime with a little bit of dedication. Plus there are even a few crops that can be put into the ground and survive the winter such as cauliflower, leeks, kale, parsnips or winter squash. With a little bit of planning, a vegetable garden can provide food for you and your family throughout every season. If you plant onions, garlic, spring onions, peas or asparagus, they can survive the winter to be ready to harvest early spring. A little bit of research will provide you with an absolute ton of variety for each and every season.

In the next chapter when we look at growing these crops, we'll look at how we go about planting, maintaining and harvesting them, as well as what techniques we can use to maximize our available space so that we can plant more. We won't be able to look at each and every type of vegetable, fruit or herb but the techniques used for growing each will be quite similar to each other. So by learning the general principles you will be able to apply them, with minor tweaks, so that you can grow whatever crops you enjoy the most.

Raising Farm Animals

Farm animals take a lot more money to start raising. For one, you need to purchase the animals themselves and it costs a lot more to purchase a living creature compared to a pack of vegetable seeds. On top of that, it also takes a lot more space to house these animals. Chicks require a coop, cows and goats should be provided a barn for shelter. You need to build a fence to keep them from roaming free. You'll also need to make sure that there aren't any harmful plants present as some animals, especially goats, can get extremely sick by eating the wrong thing.

All in all, it takes a lot of time, a lot of effort and a lot of money to start raising your own farm animals. But the rewards are well worth it. They provide plenty of food. The meat from one cow could provide food for a family throughout the entire winter. Of course, this should be combined with vegetables to keep the meal nutritious but if all you had was one cow, it could keep you alive.

But beyond the foods, which we'll look at in a moment, these animals should be seen as a long term investment. When you decide that you are going to raise livestock, you must build the necessary structures such as that barn and fence. These upfront costs are quite high but once you have managed them, you don't need to spend that much every again. If you build a barn in 2020, you could still be using that barn in the year 3000. It'll need repairs and maintenance but these costs are negligible compared to the initial investment. In that time, you could have raised hundreds if not thousands of cattle. When deciding to raise farm animals, if you are looking to profit from them then you should expect to be in the black for at least five years. It's worth noting that many people earn a profit before this but it is always better to overestimate rather than underestimate. If you can afford to lose money on your investment for the first few years, then starting won't be such a struggle. But if you need to turn a profit immediately, it is best to avoid livestock.

Of course, many of us that get into backyard homesteading don't look at it as our primary source of income. Instead, it is a way of lowering the amount of money that we spend on our food. The initial investment is much higher than we might spend on food in a year, depending on which animals we decided to raise, but projected over a ten year period we see that the amount spent on food drops drastically. The younger you can begin your backyard homestead, the more money you will save over the course of your life.

Stepping away from the financial side of things, raising your own livestock provides you with a ton of food to add variety to your dinner table. Cows provide beef and milk, the milk of which you can use to make cheese, yogurt, butter or a thousand different products. Goats provide meat and milk. Chickens provide eggs and meat. Bees offer honey, as well as beeswax which can be used to make a whole range of products as disparate as lip balm or mustache wax. This is only the most basic exploration of what these animals offer, too.

As mentioned previously, animal waste provides a source of free manure. Chicken feathers can actually be sold in bulk to people looking for craft making supplies. Bees can be rented out to farmers for help in their fields while leaving the honey for you to collect. Goats can even be rented out to serve as living lawnmowers. That's right, people legitimately hire goats to mow their lawns. Though to the goats it's not work, it's a free meal. When raising cattle, we can purchase a bull to breed our own and if that bull is strong enough then we can even rent him out to other farmers for breeding purposes. There are a lot of ways to profit from these animals, both on a monetary basis and on the basis of resources or materials.

If you are concerned about the high cost of getting into livestock farming, then I would suggest beginning first with your crops. These can quickly become profitable and this profit can be used to float the cost to get into livestock farming. From there, begin with chickens as they have the lowest starting cost of those animals which provide meat. Bees are also quite inexpensive to get started raising but they require a decent amount of space, as they like to be away from the general foot traffic of the backyard. Bees offer an interesting kind of livestock to raise, as they bolster their own numbers. We have to breed or purchase more animals when it comes to chickens, goats or cattle but a bee population will grow on its own if you allow it to. Together, chickens and bees offer the cheapest way into animal farming for the beginner.

We'll look at each of these wonderful creatures in more detail in the second half of this book. But for now, let us turn our attention towards vegetables, fruits and herbs.

BACKYARD HOMESTEAD

Chapter Summary

- Backyard homesteading is all about having a self-sufficient life, a life in which you can grow your own food and provide sustenance for yourself and your family.

- Being self-sufficient doesn't mean ignoring the outside world. We still need to pay our bills and sell our wares.

- A lot of money can be made with a backyard homestead. It takes time to learn the skills and raise the livestock or plants through to harvesting or slaughtering but it can make a tidy profit.

- Growing vegetables, fruits, herbs and grains provide us with plenty of nutrients, as well as products to sell.

- Raising livestock is more expensive than growing plants but it provides a better window for higher profits, as well as protein and iron. Depending on the species in question you can even make a lot of money by renting them out.

In the next chapter, you will learn everything you need to know about starting your own home garden. From picking and preparing the space you'll grow your crops, through to maintenance and harvesting, we dive deep into the world of vegetable, fruit and herb gardening.

CHAPTER TWO

STARTING YOUR HOME GARDEN

Starting your own home garden isn't very hard. The most time-consuming part of the process will be the beginning. Some people have it in their minds that all it takes to grow a crop of vegetables is to drop some seeds in the ground and water them from time to time. Unfortunately, it isn't this easy.

When we are starting a garden, we need to consider the physical space we are growing in. The soil needs to be checked, the temperature needs to be watched and you should know exactly how much sun a given spot is going to be getting throughout the growing season. From there we then need to adjust the soil as necessary before we can begin to grow.

Growing itself is quite easy. It is important that we maintain our crops to keep them healthy but beyond that, growing vegetables, fruits or herbs is pretty much the same experience. We could look at each one of these separately as if they were their own thing but the growing experience doesn't change. Because of this, we will spend the majority of this chapter looking at this experience before we close out by taking a look at some of the more profitable crops which you can plant, as well as those that are the easier for beginners to get started with.

Preparing a Home Garden

Before you can grow a home garden, you need to make sure that the space provides the plants with everything they'll need. Different plants require different lighting needs. For example, tomatoes need six to eight hours of sunlight but lettuce often prefers to have ten to twelve hours of sunlight. Picking the perfect spot for your garden requires you to balance various needs such as these.

To pick a spot, keep an eye on how much sunlight it gets throughout the day. If you can check it yourself, it is often a good idea to go out to the spot once an hour to take the temperature there as well. Keeping track of how much sunlight and at what temperature a spot rests will give you a much better sense of control of your garden. You will be able to pick the perfect plants that will really thrive in a given space.

But the sunlight isn't the only issue at play here. Since plants grow in soil, it's pretty important to make sure that the soil is healthy. We do this by testing samples of it with a store-bought kit to keep the nutrient levels, to see what the pH level is and to identify any harmful substances that might be present. Soil that doesn't have enough nutrients or the right pH level can be altered easily enough but soil that has harmful substances present shouldn't be used for gardening. Instead, you will either want to dig up the soil and replace it with a properly balanced soil such as you can find at any gardening center, or you will want to skip out on planting directly in the soil and instead use red garden beds for planting. We'll be looking at these in more depth in a moment.

Assuming that the soil doesn't have anything unhealthy in it, altering it can be quite easy. We can simply add some fertilizer or compost to increase the amount of nutrients present. Keep a pH level test kit on hand. If your pH level is too high then you add sulfur to the soil to lower it. If it is too low then you add limestone to raise it. Most plants want a pH level between 5.5 and 7.5. Anything about a pH level of 7.5 and the soil will be far too alkaline for the crops. Likewise, anything below 5.5 and you will find the soil to be too acidic. With that said, 5.5 and 7.5 are also pushing it at the extremes. For the best time growing, try to keep the pH level between 6.0 and 7.0.

Other concerns that you should take into consideration before planting is how much wind the space gets and how much protection it has from both the elements and the local wildlife. Oxygen is necessary for plants to survive. We try to use soil that isn't overly compact. For one, this allows water to drain better but just as important is the way that it allows oxygen to get at the roots. A little bit of wind can be useful for your plants, especially considering it makes it harder for pests to infest them, but too much wind can cause damage and wreck a crop. Similarly, it is important to water your plants and so rain can be useful but too much water will drown a crop and so sometimes it is necessary for us to provide our crops with protection from the rain. Finally, as far as protection is concerned, we should try to keep our garden away from the teeth of passersby. Deer, rabbits, mice and other animals all enjoy nibbling away at your crops and so it is important that we keep them safe from critters by using fences, raised garden beds or other techniques that make it harder for hungry critters to get at them.

When you are picking a space for your garden, keep in mind that different vegetables, fruits and herbs have different requirements that need to be met. Some want full sun, some want partial. Some are very thirsty, some like it dry. Some like a higher pH level, some like a lower pH level. For the most part, you will find that many vegetables enjoy a pH level of 6.5 and this can make it easier to group them together. But each of these requirements must be considered when you are grouping your plants. You wouldn't want to grow a plant that needs twelve hours of light in the same plot with one that requires only six to eight. You also would give that growing high pH plants alongside low pH plants will only lead to the death of one or both of these crops.

Getting your hands dirty is the best part of growing your own vegetables, fruits and herbs but it is important to take the time to plan out and prepare your garden first.

Raised Garden Beds

Raised garden beds are by far the best approach you can use to grow your garden. They will take a little bit more time and effort to set up when compared to just growing directly in the ground but they make up for it by being absolutely amazing. As the name suggests, these are garden beds that have been raised out of the ground. This is achieved by building a frame (typically wooden) that is then filled with healthy, balanced soil. Your seeds are planted in the raised bed and then the rest of the gardening experience plays out as normal. But the benefits that this approach offers are numerous.

Raised bed gardens allow us to start growing our crops earlier in the year compared to the ground. The raised height of these garden beds lets them warm up faster than the ground, so seeds can be planted in a raised bed garden before they could be planted in the ground. Another wonderful benefit is the fact that we have total control of the soil inside of the raised bed garden. We don't need to worry about harmful substances, so long as we make the bed properly. Some people have made raised garden beds out of old tires or old railway ties but these are horrible materials that should be avoided as they poison the soil around them. This means that they would poison both the soil inside of the raised bed and the soil upon which it rests. However, if we use proper materials then this isn't an issue.

When it comes to backyard homesteading, I can't recommend raised garden beds enough. They allow you to maximize your use of space. You can plant your crops closer together in a raised bed, you have fewer issues with critters and pests, you can start earlier but even cooler is the fact that you can actually design them in a ton of different ways. If you only have a thin space, you can create a long, thin raised bed. If you have a larger space, you can make it larger. If space is limited then you can even design them as a stack so that you have an elevated crop that rests above another crop. This will take some serious planning and woodworking skills to achieve, as this is a more complicated design but it is entirely achievable.

When designing your raised garden beds, keep in mind that they should be no wider than four feet. This is so you can reach every plant inside of the bed without issue. Any larger than this and you will have a harder time reaching those plants in the middle and this results in less care and attention being given them. When a plant receives less care, it makes it a risk for your garden. Disease or pests that infest this plant can go unnoticed for much longer. So while you can make your raised garden beds any length you want, don't go wider than four feet.

Raised garden beds will need a little bit of maintenance throughout the years, as the frames can crack and break and need to be replaced. However, this is made up for by the fact that the soil inside of them requires less work. Normally it is a bad idea to monocrop, which means to grow the same crop in the same field year after year. This depletes the nutrients from the soil and makes subsequent years harder to grow, often requiring strong chemical fertilizers to mitigate the negative effects. But the soil in a raised garden bed can easily have its nutrients restored. After you harvest your crops and are preparing for the winter, simply add a layer of compost followed by a layer of mulch. The compost will break down over the winter months and replace those nutrients that had been spent throughout the growing season.

Raised bed gardens are by far the most effective way to grow your vegetables and make the most out of the space available to you. We'll turn our attention now to the process of planting and growing a crop, then look at which vegetables, fruits and herbs you should consider growing as a beginner to backyard homesteading.

Growing Vegetables, Fruits and Herbs

When you decide you are going to start growing your own garden, you need to also decide if you are going to sow your seeds directly into the soil or if you are going to start them indoors first. Many people swear by starting them indoors and they have a fairly decent argument. When you start seeds indoors you can start them earlier in the year, even earlier than you could sow them in a raised garden bed. It is important that we plant our garden after the last frost of the year, as many seeds will die if they are exposed to frost. Direct sowing requires us to wait for this frost to pass. Starting indoors we can begin our plants before this frost and then transition them outside almost immediately afterward. I say almost because it requires a week of preparation to harden off plants started inside but this still result in a three or four-week head start.

However, let us assume that you are going to sow your seeds directly. Those that are planted in the ground will need to be placed into rows and carefully positioned. But this extra work doesn't need to be done with a raised bed garden. Simply sow your seeds by spreading them out in the raised garden bed and letting them come up naturally. As the seedlings grow, you will want to trim them back and remove those that are weaker. But even as you do this, a raised bed garden allows you to grow your plants much closer together than normal.

When you spread out seeds, you need to check to see whether or not that particular type of seed needs to be covered with soil or left exposed to the sun. This will determine whether or not you can simply toss them out and water them or whether you need to spend a little more time making sure they're fully covered. Either way, finish planting seeds by watering them thoroughly.

As your seeds grow into seedlings, you will thin or remove them as necessary but you can pretty much sit back for the time being. You will want to check on them once a day and stick a finger into the soil to see if the top two inches are dry or moist. If it is dry then the soil should fall from your finger when you pull it out. If it is moist then the soil will stick to your finger. Water according to the needs of the particular plants but don't overwater. Always wait for the top two inches to dry first. As a rule of thumb, it is a better idea to let your plants go a little dry compared to being overwatered. This is because overwatering can drown the plants, which in this case means that it causes the roots to start rotting and this can kill the plants quickly.

Maintaining your plants should be a daily activity. However, it doesn't need to be overly time-consuming.

Take a few minutes every day to check on each of your plants. Bring a piece of tissue paper with you, too. Look at the leaves for signs of holes or discoloration. Rub the bottom of the leaves with the tissue paper and check the tissue for blood. These are signs of infestation. If your garden is infested with pests then you'll want to blast them with water, as this knocks them loose. It is also a good idea to spray neem oil on your plants once a week, as this doesn't hurt the plants at all but it makes them taste gross to pests that would normally love nothing more than an all you can eat garden buffet. Keep an eye out for fallen plant matter and remove it whenever you see it, as pests and disease like to breed in this discarded foliage.

Fertilize the plants as needed, typically using a once weekly or biweekly application of liquid fertilizer. Some gardeners prefer to use a physical fertilizer which they only need to apply every few months or sometimes once a season. I personally enjoy a liquid fertilizer because it is easy to slow down or ramp up fertilization based on the response of the plants.

If you maintain your garden this way, you will be ready for harvest in no time. Harvesting will be determined by the species of plant you are growing. Speaking of which, let us now turn our attention to those vegetables, fruits and herbs that beginners should consider starting with.

What Plants Beginners Should Grow

So far we've looked at how we grow the plants in our garden but we've yet to explore which plants make the best fit. The recommendations in this section are based on the ease with which these plants can be grown. Along with ease, I have also tried to incorporate a range of flavors and nutrients so that even beginners can have a delicious, varied and nutritious experience from their very first harvest.

Herbs for Beginners

We're going to begin with herbs because they are among the easiest to grow. In fact, I have a hard time recommending any particular herb over the others for the sake of beginners. Herbs in general are pretty much the beginner crop. They are easier than fruits and vegetables, which makes them great for those without much experience. However, herbs should be used to accent a meal rather than replace one. So while herbs are recommended for beginners to get a sense of what it is like to grow your own garden, I still recommend that you pick either a fruit or a vegetable to grow in conjunction.

So which herbs should you start with? I am a big fan of rosemary. As far as plants go, rosemary should be considered almost a superfood. It is packed with so many nutrients as to be ridiculous. However, it is a little more temperamental than say sage. Growing sage simply requires you to plant it in a spot with plenty of sun and a quick-draining soil. Quick drainage, by the way, is one of the many benefits of raised garden beds. Another easy to grow herb is parsley. Those that plant parsley seeds directly might find that they don't really take. If this is the case, you might think that I am lying about how easy they are to grow. But the secret here is that you need to place the seeds in water before you go to bed. Let them soak overnight and then plant them in the morning. With this little trick, you'll find that your parsley takes off much easier.

Among those herbs that are easy to grow, we couldn't finish this chapter with a discussion on mint. If you are new to gardening in general and feeling a little bit nervous, try starting with mint. It will want either full sun or partial shade, which is nice as it means you can plant it anywhere that isn't full shade. Growing mint from seeds is a more difficult experience, so I recommend purchasing seedlings from your garden center and planting these. Simply remove them from the soil in their container and transplant them into the soil of your garden bed. The reason that mint needs to be referenced here is the way that it spreads. Mint loves to spread and spread and spread. It is one of those crops that will replicate itself for you. This makes it easy for beginners to get massive yields but it can also lead to the herb choking out other plants if it is planted directly in the ground or in a raised bed alongside different species. Try growing mint in its own container for the best results.

Vegetables for Beginners

While herbs are almost uniformly easy to grow, vegetables demonstrate a great range of difficulty. Some vegetables are extremely easy to grow. For example, peas and pole beans simply require you to place a trellis for support. Water them two to three times a week and they will start to grow. You can begin harvesting them daily and continue to harvest them throughout the remainder of the season. They like the soil to be 60F before planting but they demand very little of the gardener.

Lettuce, spinach and salad greens like chard and kale are among the easiest crops for beginners to grow. While we are focusing on growing in soil, it should be noted that lettuce grows especially well in a hydroponic environment and they will actually show an increased size this way. With that being said, these salad greens are easy to grow. You could harvest them all at once but if you harvest them slowly and only take what you need at any one time then you can get away with only needing to plant them once. However, because salad greens do grow quite quickly, it is possible to plant and harvest two full crops in a growing season. You may even be able to fit a third full crop if you start your seeds indoors and get them into the ground immediately following the last frost of the season. Lettuce might prove a little difficult if the temperatures get too high, as this causes it to bolt. Bolting means that the lettuce grows a flower stalk and starts to seed. At this point, it is pretty useless, though there are some tricks that can be used to get around this. But if you live in a temperate climate then you won't need to worry about bolting lettuce and should find the endeavor to be quite simple.

If you are looking to grow something more substantial then try peppers, eggplant or radishes. Each of these will be much more filling in your stomach and they will provide you with plenty of nutrients, especially if you combine them with some of the herbs that you've been growing. Radishes are a great crop to introduce you to under the soil veggies like potatoes, only they grow much faster. You could be harvesting them in as little as thirty days. Eggplant and peppers are a good way to begin growing the more complicated veggies while being easy enough that a beginner can manage them. Pick a smaller variety of eggplant (such as little finger) and stick to purple beauty or sweet chocolate peppers. These are sweeter peppers. If you want hot peppers then varieties for beginners include early jalapeno and Hungarian hot wax. Bell peppers will produce a veggie that looks more in line with what comes to your mind when you hear the word 'peppers' but they are going to be more difficult to grow. Start with an easier kind and work your way up as you feel more confident.

That last point is key and I want to take a moment to expand on it. Gardening is not a race but rather a practice in patience. When you grow anything, you need to understand that you are working on its timetable. It isn't working on yours. Your vegetables don't care if you have the last week of August booked off for a vacation, if that's when they're ready to harvest then that's when they'll be ready to harvest. Likewise, they take time to grow and you can't speed this up. Well, okay, you can but only by using chemical fertilizers which we absolutely should not be doing. So growing vegetables helps to teach you how to be patient and wait for harvest but there is another type of patience that we need to develop.

We must learn how to be patient with ourselves. Having taught people about farming and growing vegetables for years, there is a common experience which crops up again and again. Okay, you're right, I apologize for that pun. But that doesn't change the fact that new gardeners constantly throw themselves into the hardest crops. They decide to throw caution to the wind, typically with lines like "How hard could it really be?" But the truth is that it can be quite hard. Beginners don't realize the work that goes into maintaining a difficult crop. They read online about the temperature it needs and how often to water it and think that's all there is to it. Unfortunately, this isn't really the case.

Begin with an easier crop and get a sense of how vegetables grow. Bring an easy crop from seed to harvest. Experience the struggles you have along the way and remember that they are learning experiences. The more you get used to growing these easier crops, the better you are getting at growing any crop. In your first year, stick simple. In fact, you might even want to stick with simple crops in your second year, too. But once you get hands-on experience growing and harvesting, you will be much better prepared for the struggles of a harder crop.

By demonstrating patience, you won't burn out on farming the way that others have. Considering that growing your own food is a wonderful gift, I truly hope you will heed this advice and stick around.

Fruits for Beginners

Growing fruit is typically more difficult when compared to vegetables but only a little. The issue with growing fruits is that the scale of difficulty covers a wide range. Some fruits are remarkably difficult while others are no harder than growing some peppers. But you can grow some delicious fruits in a small space and enjoy yourself some natural, sugary treats in no time.

Among the easiest fruits to grow is the most widely beloved: Strawberries. Strawberries can be grown in the ground if you want but they also do well in hanging baskets. This is great because it makes it easier to maximize your available space. They love getting lots and lots of sunshine, too. If you're clever, you might be able to see how these two factors can play to your advantage. Try using hanging pots to grow strawberries so that they are in a heightened position where they get better sun. You'll notice that the pot casts a shadow and creates a patch of shade. Creatively positioned hanging pots can be used to turn an area that gets full sun into one that gets partial shade, completely changing the plants that are appropriate to grow in that space. This is perfect if you don't have a lot of coverage from foliage in your backyard.

Raspberries can be grown in containers if you have to but they will benefit from being planted in a raised bed. Unlike most vegetable crops, raspberries can be planted once but they will continue to grow year after year with a little bit of maintenance. Harvest raspberries when they are ripe and remove the branches that fruited. The remaining branches will then fruit the following year. Just repeat this each year to keep the plants healthy and productive. They'll want to be placed in the sun but even more important is their need for a quick-draining soil, which is why they are so well suited to raised bed gardens.

Blueberries and blackberries go well together because they both enjoy soil that is moist and acidic. People tend to grow blueberries in containers rather than raised beds mostly due to the length it takes them to grow. If growing from seed you can expect to take roughly three years before the plants are ready to start producing fruit. Just remember to keep an idea of the acidity of the soil, as it can be difficult to spot problems with the plants during the early stages. Often people won't realize the soil was problematic until there is a problem with the fruit. Blackberries are such a hardy plant that they can grow pretty much anywhere and they don't take up a lot of attention. If you have a spot in your garden that nothing else will grow, try planting some blackberries there. They take about a year to be ready to produce fruit, so you will know if there is an issue with the plants far sooner than you do with blueberries. But if you are planting a fruit garden then blackberries will be perfectly at home in a container or a raised bed so long as the soil remains acidic enough for them.

These four berries will make a great introduction to many of the challenges of growing your own fruits but you should also try starting with a tree. Oranges, lemons, limes, cherries and apples are all produced by a tree rather than a bush. Of these, apples are the most appropriate for a beginner. There are a ton of types but try a family apple tree. These can be purchased as dwarves if you are looking to conserve space. Dwarf trees can be grown in containers but full-sized trees should be planted directly into the ground. Skip the raised bed this time. I recommend a family apple tree because this species produces three different types of apple, which makes it perfect for those looking to maximize both their use of space and the flavors they can enjoy and bake with. Make sure that the soil is quick-draining and that the tree will get full sun. During the growing season, you will need to water it and pick fruit but otherwise, you can leave it be. You will want to prune it during winter, as this promotes more growth come spring, and you will need to pollinate the apple blossoms in spring. If you have enough space for two apple trees, you can purchase two different types and they will pollinate each other for you. This makes it much easier to grow because it reduces the amount of work you need to put in. However, if you grow bees on your homestead then you don't even need to do this much. For more about how bees help us to grow fruits and vegetables stick around for chapter eight.

Chapter Summary

- To start your own home garden you need to first set aside space for your plant beds.

- How much sun a spot gets is one of the most important features but you should also pay attention to how much shade it receives, how the wind affects it, if critters can get to it easily, whether or not it will be in the way of foot traffic and the elevation of the section.

- Soil must be tested before planting. Most of the time it is in your best interest to use raised garden beds and freshly balanced soil to grow your plants. These beds make it easier to grow plants and they will even give you a head start on the season each spring.

- Remember that different species of vegetables and fruits have different requirements in regards to pH level, watering and sun. Always research your plants first to ensure that they are grouped together in a way that won't harm them.

- Growing your own food is one of the best experiences you can have. Whether you sow seeds directly into the soil or start them inside is up to you but both have their pros and cons.

- Seeds grow into seedlings and you will be required to thin these out. This involves removing seedlings so that only the strongest are left alive to get the most nutrients.

- Fertilize plants on a weekly or biweekly basis. It is a smart idea to stick to a liquid fertilizer which allows you to take total control of their nutrients.

- Herbs are super easy to grow. Despite this, they are actually also super nutrient-dense.

- Vegetables are a little harder to grow but some easy veggies to start with are lettuce and salad greens, peppers, eggplants and radishes.

- Growing vegetables takes time to learn because you can only do so much each season. Take your time and start slow.

- Fruits are harder to grow for the most part but some easy fruits include strawberries, blueberries, blackberries and raspberries. If you want to try your hand at growing fruit from a tree then start with apples.

In the next chapter, you will learn how to grow grains on your homestead. Used to make cereals, breads and flour, these intriguing plants aren't particularly hard to grow and even a small crop can produce a large yield. However, harvesting and preparing them takes quite a bit more work when compared to their vegetable cousins. From sowing seeds to milling the harvest, the next chapter covers everything you need to start growing your own grains.

BACKYARD HOMESTEAD

CHAPTER THREE

GROWING GRAINS ON YOUR HOMESTEAD

Grains make up a huge part of our diets. From wheat to barley, rice to rye, and oats to buckwheat, there are a lot of grains in the foods that we eat. Yet growing grains doesn't seem to come up in conversations about backyard homesteading or mini-farming. This should be considered an oversight but in a way, it is representative of how widespread the myths surrounding grain cultivation are.

One of the biggest misconceptions around grain farming is that it takes tons and tons of space to get even the smallest yield. People think that grain takes thousands of feet, acre upon acre. But you can grow more than sixty pounds worth of grain in as small as a thousand square feet. It is fair to say that this takes up a lot of space if all you have is a small backyard, as the average size is about a thousand feet squared. However, nothing says that you need to grow this amount. If you have a smaller backyard then you may want to use half or a quarter to grow grains while using the rest for fruits and vegetables. If you have a larger backyard then fitting in some grains can be a great idea.

Growing your own grains allows you to make your own flour or cereals. This is a fantastic way to cut back on the groceries you pick up. You can take control of your own baking needs rather than relying on anyone else. I might not have considered growing grains to be such an important part of a backyard homestead until recently. With so many people staying at home this year, the local stores quickly ran out of flour because everyone was baking lots of goodies and trying out new recipes. This shortage saw a drastic change in the diets of my friends and family but thankfully I can expect a harvest in the near future and this will help me to provide homegrown flours to them. So while growing grains is great for saving money, the way it enables you to separate yourself from the grocery stores' chain of supply and demand is even better.

This ties into the second widespread misconception around growing your own grains, though only indirectly. It seems that people have it in their minds that growing grains requires a large investment in equipment and specialized machinery. How this misconception came to be is rather confusing. I assume that it has to do with the widespread imagery of farm equipment harvesting wheat and other grains, though this is an assumption and I can't guarantee it to be the truth. Historically, grains have been harvested with scythes for generations. Or you can use a pair of shears. To harvest grains, we cut down the stalk but then we need to thresh it, which is simply the term we use for removing the individual grains from the seedheads. This is achieved by simply whacking the stalks with a stick until they all come off. The grain then needs to go through a process known as winnowing to remove the chaff. This is done with a fan but it doesn't need to be a specialized fan, a $10 fan from your local Walmart will get the job done. Finally, a blender can be used in the place of a mill to convert the grains into flour. If you bought each item purposefully for their use in growing grains, you are looking at around $50 with the most expensive items being the blender and the fan. Note, the two most expensive items are also the two that can be used for countless other activities around the house and they are also the equipment that people are most likely to own already.

So growing grain take up less space than people think and it costs less than people assume it will. If you want to have a backyard homestead that meets all of your needs, growing grains is an absolute necessity. We'll look at how to grow grains in a general sense since they do really follow in each other's footsteps. From there we'll turn our attention over to the needs of specific grains and how to harvest and prepare them.

How to Grow Grains

Growing your own grains is surprisingly simple. However, there will always be unique challenges that arise based on the species of plant in question. Growing wheat will be a little bit different from growing oats and so on. Yet the process is primarily the same. In this section we are going to look at this process so that we can see what it is to take grain from seed to harvest. Afterward, we will look at individual grains to get a better understanding of how they differ. This is primarily due to factors such as how much to water them or how long they take to grow, when to plant them. These should be considered addendums to this simplified process.

To begin with, we need to pick out our seeds. These are simple enough to find. Your local garden center should have plenty and if they don't then you can always find them online. While they should be rather inexpensive, the cost is going to be determined by the species in question and the amount you need to purchase for your backyard homestead. If you are only growing a small patch then you only need a small order, though you always want to purchase and sow more seeds than you expect to harvest. Keep in mind that not grains are planted at the same time. Speaking only of wheat, there are both winter and spring varieties that are planted in autumn or spring, respectively. Since you can't expect to plant different types of one plant at the same time, it would be ridiculous to think that what works for one species also does for another. Make sure that you check the seeds you are planning to buy to make sure that they are appropriate for the season you are planting.

Even among just winter wheats alone, there is a distinction between soft and hard varieties. A softer wheat has a small amount of gluten that makes them appropriate for those who have a sensitivity. Hard wheats have a high gluten content and are more regularly used for breads and pastas. Picking between hard and soft varieties of grains shouldn't have a large impact on growing them, it's more important when it comes to using and consuming them. I recommend sticking with a soft grain whenever possible, as the lower levels of gluten will make it easier to cook for and gift food to friends and family. However, this is primarily a judgment for your taste buds rather than any difficulty you'll have growing them as a beginner.

Assuming that you have purchased the right seeds, you will be planting either in spring or autumn. While it is extremely important to get the time of planting right, the actual act thereof is much the same. We judge the right time to plant-based on the weather. With spring-planted seeds we want them in the ground after the last frost of the winter season. Autumn seeds are planted just before the first frost. This results in a harvesting season either in the middle of May or in the autumn just before that frost hits.

Before you can plant the seeds you need to test the soil. I recommend getting a kit and seeing if there are any heavy metals or harmful chemicals in the soil. If there are, it doesn't necessarily mean that you can't use that spot but rather it tells us that we need to prepare it first. We do this by digging up a couple feet of the soil, laying down some tarps and then filling it back in with healthy soil that we purchased or made ourselves. For beginners, I recommend purchasing a pre-made soil that is already rich in nutrients. If you are using your own or a soil that hasn't been mixed then you should add some compost to it to boost the nutrients present. Keep in mind that we shouldn't just get soil from the Earth directly and expect this to be fine. Proper gardening soil is treated beforehand to kill off any harmful microorganisms present and this goes beyond the scope of our discussion. So stick with a store-bought soil and add compost if it hasn't been mixed in already. You should still use a properly formulated soil for growing your grains even if your test didn't reveal any harmful chemicals but you can skip the tarp.

Fill in the garden bed with the soil. While some crops will have you creating little hills of soil, such as we do with potatoes for example, grains grow best when the field is even. You don't need to worry if there is a slight lean to the soil or if it isn't perfectly leveled. So long as it is mostly even or appears to be then it will be fine. This is done to create the best possible growing environment we can but mother nature is never perfectly even herself. Rake the soil or pat it down with your shovel until it looks to be even and then check this step off your to-do list.

With the soil ready, it is now time to get sowing your grain seeds. Sprinkle the seeds liberally. We do this for two reasons. The first is a fact shared across all plants, a universal truth when it comes to our green friends. Just because you have a seed, that doesn't guarantee it will grow. Some seeds are sterile. While we don't have a good way of testing our grain seeds this way, harder seeds that need to go through a soaking process demonstrate this fact quite clearly. Seeds that are soaked in water before planting should sink to the bottom of the liquid after a little while. Those that continue to float are sterile and tossed out. Unfortunately, we can't run a test like this so we must sow our seeds with the assumption that a good number of them will simply not germinate.

The other thing that grains have going for them is that they can and do grow much closer together than our vegetables will. When we plant our vegetables, we follow one of two paths. The first is to carefully position them and then sow three or four seeds in each location where we have planned for the grown plant to be. The other technique, which we discussed in relation to raised garden beds, is to sow the seeds at will and let them land where they want. Regardless of which of these we followed, there is a second step during the seedling stage of our plants' life cycle. With vegetable crops, we thin out the seedlings by removing those that appear to be weaker. This allows the stronger seedlings to get better access to the nutrients in the soil since they are no longer competing for them. But with our grains, we like to grow them especially close together. So when we sow our seeds, we are expecting them to grow together as tightly as blades of grass do. This is useful for maximizing our space, something which we will continue to address and aim to achieve throughout the book.

Once you have sown the seeds, take your rake and use it to lightly till the soil. There is no need to build up the soil into mounds of any sort, so don't go crazy with the rake. Simply go through and use it to cover up the seeds. Water then thoroughly. This is extremely important but we'll discuss watering more in just a minute. First, we must talk a little more about covering our seeds. We rake the soil to cover them that way but this isn't enough. During this stage in their life, the seeds are extremely vulnerable. There are many different species of birds that feed on seeds and they would love nothing more than to have a free meal. If you don't protect your seeds, then you shouldn't be surprised when they are eaten. The problem here is that you want to protect them in such a way that still lets them soak up the sunshine. If you can get a mesh tarp or tent then this will do the trick. The holes in the mesh will allow plenty of sunlight in while preventing birds from getting easy access. It isn't foolproof, determined birds can learn how to get under it or just straight up chew their way through, but it will take care of 99% of the issue. Really can't ask for much more than that.

When it comes to watering, the species you are growing are going to determine whether they need a lot of water or a little. But pretty much every grain likes to get lots of water while they are trying to germinate. Water seeds once a day to ensure that there is enough moisture in the soil to support them. It is best to do this earlier in the day, before noon if you can. This gives the water enough time to evaporate before the night. This isn't as important with grain seeds as it is with vegetables and fruit but it is always a good idea to build a habit of only watering your plants before noon. This simple habit shouldn't be too hard to integrate into your life and it should keep your garden from suffering from root rot and other harmful issues that arise from overwatering.

As your grain grows, it will start to come into its color and you will notice that it gets much heavier. This weight is due to the grains coming in. Since the grain grows on the head of the stalk, you should notice that they begin to bend down a bit. This tells you that it is time to harvest, or at least that it is extremely close to time to harvest.

We'll talk about harvesting in just a moment, first, let's take a look at a few grains to get a sense of their own unique and specific needs.

Grains for Beginners

Of those that are recommended for beginners, the easiest is wheat. In fact, the guide above can be taken as-is for wheat. You will want to water your wheat three times a week after it leaves the seed phase of its life cycle. But other than this, just follow the previous guidelines and you'll be enjoying wheat in no time. Whether you plant in the spring or the autumn is determined by the type of wheat you choose.

Rye is similar to wheat, though you want to make sure that you pack it in tightly. Instead of just using a rake to till the soil after sowing, it is recommended that you then pack the soil. This is done to ensure the maximum amount of contact between the seed and the soil, something that is needed to properly germinate. Rye is grown in the autumn, with the window for planting being a fair bit larger than other plants. While it is a good idea to plant rye a couple of weeks before the first frost, you can get away with planting it up to a month after as well. In fact, while these dates make for the best crop, you can get away with planting pretty much anywhere from August to March. The stuff is very hard to kill, though out of season planting does reduce the flavor. Rye enjoys wet soil and if the winter is mild then it will grow quite quickly. Quick growing rye is cut to maintain an even height but rye in poor soil or a harsh winter can be left to grow on its own, simply water it and harvest it at the appropriate time.

Oats are another staple grain and, while they prove a little harder than wheat and rye, they are a good fit for beginners. Oats like quick-draining soil and they yield the largest amounts of product when they're planted around mid-May. Unlike wheat, oats should be grown in rows that are between half a foot to a foot apart from each other. Oats are particular about the amount of nitrogen in their soil. Too much nitrogen and too little nitrogen both cause issues with the harvest. The more water present, the more nitrogen is needed. The biggest issue that oats tend to have is a deficiency of manganese, one of the micronutrients which plants require to thrive. Oats are a great crop if you find that you are having issues with pests. Most insects don't want to eat oats and so they work well as a crop you can ensure will make it to the harvest. Wait until the hulls have turned from green to tan before harvesting. The youngest kernels on each piece should be a cream color. Harvest when these conditions are met, as leaving them to mature too long will result in a lower quality yield, as well as increase the chances that a storm or strong wind breaks the stems and reduces the size of the yield alongside the quality.

How to Harvest, Process and Store Grains

Grains mature quickly. Some go from seed to harvest in as little as a month. You should know how to harvest them before you plant them, as it is easy to put off learning until it's too late and they've been left in the ground for too long. But harvesting grains follows the same pattern of being quite simple.

Watch the stalks as they change color. They'll go from green to brown. When they do, you know it is time to start harvesting. Grab your scythe or whatever cutting tool you are comfortable using and start bringing them down. Make your cuts just above the ground, about an inch up on the stalk.

When you have enough stalks cut down to make a bundle, tie them together. You don't need to use a thick rope, some twine should do the trick. For the next two weeks you will want to keep them out of the rain so that they can dry. You can toss them on a floor in your shed or you can hang wire and attach them to it. Many people prefer this latter technique because it elevates the stalks so that gravity can help them to dry out by pulling the moisture down. Hanging them also prevents them from being laid in a puddle of water on the floor. It is unlikely that the stalks contained enough moisture to form puddles, it is not impossible.

It takes roughly two weeks for grains to dry out. You will be able to tell when it is ready by its texture. The grains should be hard to the touch. Pop one in your mouth and chew it. If it is crunchy then you know it is ready for the next step.

Lay down a tarp or some other fabric. This is laid out to catch all of the debris that we'll be creating. Toss a bundle of grains onto the tarp, grab a stick (or a dowel or a light hammer, pretty much anything you can use as a blunt weapon) and start whacking away at them. This is the threshing stage. We abuse our grain stalks to knock the grains off of them. Once a stalk has released all of its grains, you can toss it into your compost pile. Collect the grains in between each bundle.

Alongside the bucket or container you use to scoop up the freshly threshed grains, you are also going to need a second container and a rotary fan. Place the fan on a medium breeze and point it at your second (empty) container. Take a handful of grains and drop it into the container, making sure that you hold them high enough that the breeze can catch them as they drop. Grains have a paper-like layer on them that we call the chaff. You could eat it if you wanted to but it wouldn't be very tasty. It's meant to protect the grains and so we no longer need it at this stage. Since this layer is extremely light in weight, the fan should be powerful enough to blow it away. After dropping the first couple handfuls, check the grains in the second container to see if the chaff is still attached. If it is then the fan should be set a little higher. The removal of grain chaff is referred to as winnowing.

The grain is ready to be stored once the chaff has been removed. Use glass jars and store them in a cool dark place. Exposing the grain to light can promote the growth of molds and fungus. If you notice any discoloration in your grains, or if you spot a substance that looks out of place, then your best bet is to toss that jar. Chances are it either wasn't sealed properly, it was stored in too warm of a place or it was exposed to direct sunlight.

There is one step left for your grain and that is to mill it. This used to be done in mills that needed to be manually worked. This was achieved both by human labor and by animal labor. As technology advanced, systems were designed for automated mills that used water or wind in place of labor. Nowadays we don't even need to worry about that. Just toss your grains into a blender and give them a good beating. Make sure that the blender is made to withstand heavier objects, as weaker glass blenders can be broken this way. Whenever you need to mill grain for some flower or the likes, just open up your jar and toss some in. For wheat and rye this will be the last step but if you are looking to mill rice or oats then there is actually one more step.

Oats, rice and buckwheat are just a couple of the grains which can't be milled right away. They have a hard hull. This is like a skin on the outside of the grain, only it can be extremely tough and so it needs to be removed before milling. One technique is to put these grains into a blender and give them a quick beating. This isn't enough to mill them but it can break open the hull. Some grains will naturally separate from the hull this way but others will crack and then need to be removed by hand. If you purchase an at-home grain mill, a specialized appliance that should replace your blender if you get one, then you should be able to get an attachment which will break the hulls for you. Some kitchen counter mills will even come with the feature as a selling point.

And there you go. From seed to harvest, threshing and winnowing, all the way to cracking the hulls and blending your grains, you now have what it takes to grow grain as a part of your backyard homestead. I highly recommend that you give it a try and see just how satisfying it is to bake a loaf of bread from ingredients that you've grown yourself.

Chapter Summary

- Grains are a great crop to grow for cereals and flours. They can be stored for a fairly decent length of time, they don't take up very much space and they grow quickly.

- Plus you can always sell them, though they won't earn as much as a crop of veggies or fruits will.

- There is an idea that growing grain takes a lot of expensive equipment but this simply isn't true. A knife and a stick are all you need to process your grains.

- Grains are grown from seed, which is typically sowed quite liberally in an area. Many grains can be grown in tight clusters compared to vegetables or fruits. They're more in line with herbs this way.

- There are winter and spring varieties of most grains that are sown in May or September. They also come in hard and soft varieties with hard grains having a much higher gluten content than soft grains do.

- Water your grains based on the needs of the species in question, this will require you to do some light Googling.

- Grain will start to change color from green to yellow and the tops will get heavier, causing a slight lean, this tells you it is time to harvest.

- Beginners should try starting with a species of wheat, rye or oats. Wheat is the easiest of the three but it is quite versatile.

- Grains are harvested when they have changed color and show a lean. They are cut with a sickle and then left to dry.

- Dried grains are then beaten with a stick to knock them off the shaft.

- A fan is turned on and grains and dropped in front of it to have the chaff blown off.

- You can then store or mill your grains, though some grains need to have their hull cracked before they can be milled.

In the next chapter, you will learn how to set up your own barn. Whether you need to make one from scratch, convert an existing shed or purchase a prefabricated shed to use as a barn, you can find out how next.

CHAPTER FOUR

SETTING UP YOUR BARN

This chapter could be thought of as the bridge between the first and the second section of this book. So far we have spent our time exploring the horticultural side of mini-farming. We've grown vegetables, fruits, herbs and grains. We're now going to move into raising our own animals for meat, eggs, milk, soap, fiber and honey. But raising your own livestock is going to take a lot more work than planting a garden does. Specifically, raising your own livestock requires a much greater initial investment.

This investment needs to cover quite a few costs. There is the cost of the animals themselves but that is only the beginning. We need to be able to provide them with food, water and shelter. A fence must be erected to keep them from wandering all across the countryside. A fence will also offer a level of protection from predators. Each of these steps requires money, time or effort, and some of them need all three.

In this chapter, we are going to tackle the biggest of their needs: the barn. But we won't just look at a barn as something that exists in a vacuum. We'll also consider how the barn's placement directly affects the placement of the fence. These two objects should be considered inseparable from each other.

We're going to look at setting up a barn from two different angles. The first angle is how to build your own barn. As you'll soon see, this is an extremely time-consuming process that requires an understanding of carpentry and architecture. This option will give you the largest sense of control of your barn but it takes a long time and it can end up costing you a lot of money. The other angle we'll approach this topic from is the converting of a shed into a barn. Specifically, we'll look at how to convert a prefabricated shed into a barn. If you already have a shed then you should be able to apply the same techniques to your shed as we will to the prefabricated one in our discussion. With our barn settled and out of the way, the remaining chapters will deal with raising livestock and the different commodities they produce.

Building Your Own Barn

There are three stages to building a barn, each with a number of steps that must be followed. In no way should this process be considered easy, even if many internet blogs claim that it is. Rather, it should be acknowledged upfront that it can be quite difficult. But it is not impossible. It will take time and effort but it is absolutely possible. If you have carpentry skills then it will be easier for you than someone without, though you could always hire a professional to help you out.

The three stages of barn construction are to choose a location, build a foundation and, finally, to build the barn itself. These could be thought of as planning, preparation and actualization. Typically, when it comes to mini-farming, planning is the longest step in any process but here it will be the shortest. There is a lot of labor that goes into building a barn and this only becomes clear in the second and third stages. Before you undergo construction, consider your own physical health for a minute. Are you going to be able to dig, pour, hammer and raise the sides of a barn? If you have medical conditions that prohibit strenuous work then your best bet is to convert a prefabricated shed like we'll be outlining shortly.

If you are in good enough health to continue then it is time for step one. Unfortunately, this is a pretty boring step but a necessary one. If you want to build a barn, you must first find out if you are even allowed to. This means you need to find the building codes for your local area. If you are in the United States then this can be found on the government's Codes Enforcement website. Those in other countries will need to find these for themselves but nearly every first and second world country has them listed online at this point. If you Google "Building codes" plus your local area, they should come up as the first or second link. These codes are going to be a slog to read through but they will let you know everything you need to start building. If you need to get government permission to erect a barn, what size is acceptable, what regulations need to be followed for plumbing or electrical wiring, all of these and more will be listed in these codes. Make sure that you follow them to a tee, otherwise, you can end up on the wrong side of a legal battle that will quickly dash your dreams of having your own backyard homestead.

Once you know the legality of your barn, it is time to start picking the space. The right space will depend on several factors: the ground it is laid on; the patterns of the wind; the amount of sun and shade; and how much space you have available in the first place. Each of these has a preferred state but it is hard to find a spot that is perfect for all of them. As you read this guide, take into consideration which features are absolute necessities for you and which you can get around. With a little creativity, you should be able to make your new barn work fantastic even if you have issues with the wind or the sun.

The ground you choose will be the most important piece of the puzzle to get correct. You want the ground to be level, with a densely packed Earth rather than an overly sandy structure. Rainy areas will benefit from fast-draining soil but too much sand or too much gravel will result in a weaker foundation. While it must be flat, try to avoid picking a spot that is at the bottom of a hill. Water will flow down the hill when it rains and this can lead to unexpected flooding. When picking this spot, remember that you need to set aside space for two different buildings, in a way. The first is the barn itself. This definitely requires flat, secure land. But you also need to consider the fence and the size of the pasture the barn will open into. You want your animals to have enough space to stretch their legs, for one, but you also should have a section for feeding, a section for relaxation and a section for breeding (if you are going to breed your own animals, which I recommend). The pasture that this fence outlines doesn't need to stand up to the same rigorous demands as the space where your barn will rest.

Pay attention to the wind patterns. If you can, you are going to want to situate your new barn downwind from your house. This is done to avoid carrying the smell of manure and animals into your home. Some places will find that the wind seems to come from all directions. This makes it hard to pick the perfect spot but if you design your barn with doors on all sides then it will give you the ability to control the ventilation and reduce the smells around the house. Similarly, you should also pay attention to the sun. The best placement will have partial sun, preferably with the shade being cast on the spot around noon when the temperatures would be the highest. This will help to prevent issues in your animals such as heatstroke. With these taken into consideration, you should have your spot.

The next stage is to get the foundation down. We do this by first excavating the site where the barn will be. We dig up roughly half a foot of dirt in the shape and size that the final building is going to be. As this is a rather large dig, you will find that it works best with an excavator or even a bulldozer if you can afford to hire or rent one. Place all of the soil you've dug up off to the side, it will prove useful if you ever take on other landscaping tasks. You can even make a little bit of money off of it by selling it to others. While you're doing this digging, make sure to keep an eye out for rocks and roots. You'll want to remove these now so they don't weaken the foundation in the future.

The next step is more digging. But this time we dig a trench around the perimeter. The depth and width of this trench will be determined by your local building codes. This hole is called the footer trench and we will be filling it with rebar. The goal is to use the rebar to create support and tension against the sides of the trench. Do this the whole way around. This will result in a skeletal-like appearance. We then fill in the trench with crushed stones. These can be purchased wholesale or bought at your local hardware store in weighted bags. Fill in the trench so that the stones leave only half a foot to a foot of space from the top of the hole. Use a shovel to even out the stones. The flatter you can make the surface of these rocks, the better.

Once your rocks are in place, it is time to fill the trench with concrete. You want this concrete to completely fill the trench, leaving only a few inches from the top. Again, we want to try to create an even surface with this. After your first pour it will take roughly three days for it to harden fully. When it has taken some more of your rebar and places it along the perimeter. This will offer more support for our next pour, which we will undergo now. We don't need to bother adding more stones this time around, so simply pour the new concrete up to the top of the trench. This also takes three days to set. After the trench is complete, lay down some wire mesh to the dimensions the building will be and then pour half a foot of concrete. This will fill the hole you originally excavated. All together you now have a perimeter of solid concrete which is firmly held in place by the Earth itself and you have concrete flooring to serve as your foundation.

Take note that concrete flooring is quite uncomfortable for your animals' feet. You will need to make sure to lay down mats and hay. Pretty much anything you can do to soften the experience for your animals will result in healthier and happier livestock.

Finally, it is time to build the structure itself. This is done by first constructing a stick frame. This is a frame made from 2 by 6 studs that outlines the walls. This is then followed by a frame for the roof. This can be flat or sloped. Which is right for your barn will depend on what you think looks best and what the weather's like in your area. If you are in an area that gets lots of rain then a sloped roof is best because it prevents water from pooling up. This may sound like a minor concern but water does have a weight and too much of it will damage the integrity of the roof.

With the shape of the barn in place, next you will want to install the roof itself. Lay down half an inch of plywood. This is then followed by some roofing felt and finally by some metal sheeting. The walls come next. Your best bet is to go with a pre-cut wood siding such a board & batten siding. These can be purchased in your required dimensions to save time cutting and sawing. All you need to do is nail them securely into place. Remember to leave enough space for a couple of sliding doors. You may also want to cut out holes for a couple of windows to improve airflow. Make sure to attach shutters so you can close the windows up during storms, or the barn will get too wet and it will be impossible to keep it properly heated.

There you have it, a step by step guide to making your own barn. You'll want to divide up the inside into stables but these will be determined by what type of livestock you decide to raise. If you have limited space for your pasture, consider setting aside a space inside of the barn for feeding. Make sure to include plenty of hay for bedding, too. At first, you will probably find that the barn is rather empty and plain. As you get more involved with your animals, you will get a sense of what tools you need and how you can best support them. Within a matter of a couple years, your new barn will go from feeling empty to feeling full of life and thoroughly lived in and this is one of the coolest things you can ever experience.

But building a barn isn't for everyone. It can be quite tiring and it can take quite a long time, too. The benefit of building your own barn is the level of security you give the structure through the concrete foundation and the way it lets you fully take control of its dimensions. But if you are looking for a quicker, easier way to get started with a barn of your own then you'll want to stick to prefabricated buildings. Let's turn now to see what this entails.

Converting a Prefabricated Shed into a Barn

Most sheds are going to be too small to house cattle. Cows take up a lot of space and unless you only have one or two, you will need to build your own barn or to string together several prefabricated sheds. But turning a shed into a barn for goats is quite easy and the sizes work out far better. Don't take this to mean that you can't turn a shed into a cattle barn, just that it is far less likely and requires far more work to do so. For cattle, you are best sticking to the previous guide on constructing your own. Also, as mentioned previously, this section is primarily concerned with prefabricated sheds but it works for converting an already existing shed into a barn. The focus on prefabricated sheds is simply to show that those without a shed can still easily get one set in place and prepared for their animals.

A prefabricated shed is a great choice for those that want a simple time making their barn. Seek out a local retailer and purchase from them. They should have some that are immediately ready to be placed but most retailers allow you to pick from an array of styles, sizes, colors and the like. Going with an already prepared one will get you a barn quicker but you will be stuck having to settle for what is available. But you shouldn't be in a rush at this stage. After all, we haven't purchased any animals yet because we don't have a space dedicated for them. Spending a couple of weeks to get a prefabricated shed that best fits your needs should be a breeze. I fullheartedly recommend having patience and perfecting your shed rather than rushing into a purchase. Remember, too, that zoning laws should be consulted before purchase.

Prefabricated sheds don't require you to construct a concrete foundation. However, you do still want to create some sort of foundation. Crushed stone is the best choice as it will help to improve the drainage. In the long run, this will reduce the amount of water damage and create a much more secure placement. Dig a hole that matches the specifications of your prefabricated shed. This only needs to be two or three inches deep. Once it is dug, fill it back in with crushed stone. There you go, that's all it takes for the foundation of a prefabricated building. You shouldn't need to worry about attaching doors or windows or anything of the like. These can all be given to the manufacturer who will put them in place for you.

Guess what? The shed will be delivered to you and often the company will even set it up where you specify. This means that the outside aspects of your shed are handled. Now you will need to fill the inside up. How you divide the space will be determined by how many animals you own and what you are looking to do with them. You might use a part of the shed as a milking station while you place stables and lay hay on the other side. Whatever you do, you should aim to have a plan in place before purchase. Since the purchasing and laying of your prefabricated shed are as easy as contacting the manufacturer, let us turn our attention from building over to the layout.

Start by figuring out how much space you need in your barn. List all of the animals you will be raising in it, how much equipment you need to store, how much space you need for milking or birthing or whatever else you plan to do. Once you have this in place, take a ruler and a piece of paper and start designing a mock-up of your barn. If you are in doubt about the size of any particular component, always settle on estimating it will be larger and take up more space than you think it will in reality. It is always, always better to have more space than you need. More space allows you to grow, too little space cripples your aspirations. Keep in mind that the manufacturer will have an upper limit on the size of the building. You should use this as a guide for the maximum amount of space you can allocate in your mock-up.

Some features which you will want to consider are haylofts, sliding doors, overhangs and pressure-treated floors. A hayloft gives you a place to store hay so it doesn't get in the way and it is easy to add to a shed whether or not the manufacturer includes them as an option. Sliding doors are pretty self-explanatory. You don't necessarily need them, as swinging doors with a wide arc will suffice but sliding doors are the easiest to open and close quickly since the door stays against the side of the barn. Overhangs in front of the doors are smart additions because they offer better shelter from rain and snow. While most manufacturers will have overhang options, they are extremely easy to build yourself. Finally, pressure-treated floors may have fewer issues with rotting but there is some debate as to whether or not they expose the livestock to harmful or toxic chemicals. Whenever there is a debate of this type it is best to avoid the feature. By avoiding the feature, you avoid the risk entirely.

If you are raising goats then there is one other feature that you will want to add yourself. As cute as they are, goats just love smashing things with their heads. Each other, fence posts and even the walls of their barn. If you let them bash the walls long enough then they will eventually break them. Reinforce the walls by adding a few sheets of plywood to the inside and you may even want to consider adding some padding to the plywood to make your goats' headaches a little kinder. This will also serve as insulation to help your animals stay warm during the colder months.

A prefabricated shed won't come with electricity or plumbing so you will need to install these yourself if you must have them. Personally, I don't think that electricity is necessary. Purchase a portable, battery-powered work-light from your local hardware store. This will provide plenty of light while you're working in the barn. Plumbing is always nice but you can run a garden hose in through a window easily enough. This will keep your building simple, plus it has the added benefit of keeping costs low.

There are a lot more additions you could make to improve your barn and make it easier for working with your animals but I want to avoid just listing off a bunch of things which you may not want or even realize the purpose of them. Rather, I want to stress the importance of viewing your backyard homestead as a learning process. It isn't recommended that you plant every type of vegetable, raise every type of livestock or purchase every possible tool when you are just starting out. This isn't a race, it is a marathon. Your homestead should be seen as a way of supporting you and your family for years to come, even possibly generations to come. You will be adding to it, experimenting with it and fine-tuning it for the rest of your life. This is one of the things that makes it such an enjoyable experience: it is constantly evolving and you are constantly learning more and getting better at it. Start out with what you need to begin with and grow from there. It'll make the experience more enjoyable, less expensive and less stressful and that's just good living.

BACKYARD HOMESTEAD

Chapter Summary

- Setting up a barn is a necessary step if you want to raise livestock ethically. While a barn can be converted into a barn for smaller animals such as goats, cattle will need a larger structure depending on how many of them you have.

- A barn is built by first digging out the foundation. A trench is dug along where the walls will stand. This is filled with rock and metal and concrete to create a solid foundation. The floor is then filled in with concrete as well.

- With the foundation in place, walls are built followed by the roof.

- Look up the legality of raising a barn in your local area, not everywhere allows you to raise a structure without dealing with some serious paperwork.

- You want the land you build on to be even and to have some level of wind protection. You will also want to build your barn downwind from you so that it doesn't stink up your home.

- If you have a shed already then you will have an easier time-converting this into a barn than you would building a whole new one. If you don't have a shed, you could always buy a prefabricated one and convert this into a barn easily.

- A prefabricated shed can be brought directly from the supplier but often they allow you to specify your needs. Doing so takes a little longer for them to put together a shed for you but it is the best choice for converting into a barn. Get a sliding door and a couple of windows in yours.

- You don't need to create a concrete foundation for a prefabricated shed, which makes it a much quicker setup. Often the company you purchase from will set it up for you.

- A prefabricated shed won't have electricity or plumbing but you shouldn't need it.

In the next chapter, you will learn how to raise chickens. These creatures are easy to tend to, though they'll need a fence and a chicken coop. If you can provide those and a little bit of time then raising chickens should be no problem.

CHAPTER FIVE

RAISING CHICKENS

Chickens are among the easier livestock to raise and farm. They don't take up a lot of space, they aren't overly demanding and they are fairly inexpensive to get started with. You will need to build a chicken coop (or convert a shed into one) and we'll cover how to do this here. But beyond the coop and a fence, there isn't much that you need to do. You'll want to feed them and provide them with plenty of water, plus you'll want to clean up after them. If you can manage these simple tasks, you should be able to manage chickens.

Before you go any further, you need to figure out two things. The first is whether or not you are even allowed to raise chickens on your property. This is another of those restrictions that you can find by searching out your local zoning laws. If you are allowed to raise chickens then you need to figure out what breed you want. There are egg-laying breeds, breeds that are grown for their meat or even some breeds that are raised for both purposes. To get a sense of which breed is right for you, you need to consider the three benefits of raising chickens. We'll look at these first, turn our attention over to chicken coops and then learn how surprisingly easy it is to take care of these birds.

The Benefits of Raising Chickens

Chickens are quite productive birds considering the fact that they don't really do much of anything. They just wander around, pecking at their food, enjoying a drink. They aren't work animals, so you can't use them to help around the farm. And they're certainly not useful for transportation. Just try to picture a chicken-drawn cart without chuckling.

But despite this, chickens are one of the more profitable animals you can raise. One reason for this is their low cost to get started. While it will depend on the breed in question, you should expect to pay no more than $30 for a live chicken. This price will often be as low as $3, though this is typically a sign of a less attractive breed. When planning to raise your own chickens, assume that you will pay $30 for each one. This way you can come in under your budget and enjoy the savings.

As said, chickens quickly pay for themselves when you weigh up the benefits of raising them. While patience and the ability to extend compassion to an animal might be considered a benefit, we're going to stick with those that have a direct impact on your pocketbook, your diet or your backyard homestead as a whole. This leaves us with three key benefits to consider.

Meat: Growing chickens for meat is a way to get paid more, faster. Meat birds are typically selected for their size and the speed at which they grow. Some breeds are ready to be sold to the slaughterhouse in as little as six to eight weeks. This level of turnaround makes them a more profitable venture than egg-laying hens but it also means that you need to be okay with spending weeks with a bird that you are planning to kill. For many, this just simply isn't easy to do. However, it is actually crueler to let meat chickens continue to live. Their bodies aren't designed for long-term functioning and they will suffer from heart attacks and organ failure after a couple of months.

The price you can get for chicken meat is calculated by the pound but it can fluctuate widely depending on the local competition and other factors of the marketplace. In general, you shouldn't accept less than $3.50 a pound. If you are lucky you can earn as much as $6 a pound. Most chickens will weigh between five to ten pounds so you're looking at anywhere from $17.50 to $60 a chicken.

Eggs: For those that don't want to deal in quick turnarounds, or those who don't enjoy the thought of condemning a creature to die, chickens can be farmed for their eggs. America's favorite breakfast food, eggs are a pretty profitable venture. A carton of a dozen chicken eggs will sell for as low as $2.50 or as high as $5. If you raise half a dozen chickens then you can expect them to eat about $30 of feed a month, so this number needs to be compared against the cost of the chickens and the cost of feeding them.

Assuming that you spend $30 on each chicken, you are looking at a cost of $180 to start and $30 a week after that. Chickens don't normally lay eggs during the winter months but you can purchase a bright light for their coop that will trick their biology into thinking it is still laying season. Assuming you do this, we can calculate how many eggs a chicken lays in a year. On average, chickens lay five eggs a week. So six chickens each laying five eggs a week gives you 30 eggs a week or 1560 eggs a year. This equals 130 cartons to be sold in the year, which means you can make anywhere between $325 and $780. On the low end, this will not cover the cost of feeding in the year but on the high end, it covers double the cost.

But the best part of this is that you don't actually need to sell them all. Just by raising a couple of chickens you can remove eggs from your grocery list. If you purchase more than $30 of eggs in a month then you'll be saving yourself money in the long end. Plus you can still sell whatever leftovers you can't eat. This makes farming chickens for eggs a highly attractive venture.

Manure: Finally, the third benefit is the chicken's poop. That might sound a little gross but farmers pay good money for manure. It is high in the nutrients that plants need to thrive. And, trust me on this one, you are going to be dealing with a lot of it. One of the worst aspects of taking care of chickens is cleaning up their coop to remove all of the excrement. Yet if you are running your own backyard homestead then you will have a use for all of this crap. It turns the biggest negative of chickens into a major positive. That's pretty awesome any way you cut it.

You will quickly find that they are producing more manure than you can use. Don't just toss out the extra. Get yourself some bins for storage and store the extra, you can give it as a gift to your fellow gardeners or sell it to local farmers. Just make sure that you place the storage bins far enough away and downwind so you don't need to suffer through the smell every day.

The Perfect Chicken Coop

Building a chicken coop is a lot easier than you might think. It doesn't need to be as fancy as a barn, nor does it need to be as big. In fact, you could give a chicken a large box and they would be perfectly content with it. Of course this box wouldn't fill the chicken's needs and it would be safe to assume that your chicken's life expectancy would plummet. But it does point towards how little chickens care about where they live. We should care, though, as we can consider issues like predators and the weather where they can't. To this end we aim to achieve a balance between several requirements.

But how we get that balance is entirely up to you. You could choose to make a large coop or a small coop, you could make it square, rectangle, triangle or any other shape your mind can imagine. So long as you can provide these basic needs, you will have healthy and happy chickens on your hands. It should also be noted that you can put together a chicken coop for well under $500. In fact, you could make one for closer to $100 if you are mindful of the materials and the size. But rather than slow the book down to build a coop, let's look at the requirements our chickens need.

A coop first and foremost offers your chickens shelter. If the wind is blowing too strong they can go inside and rest. If it is raining then they don't need to go out and get wet. But speaking of rain, you need to ensure that this shelter actually works. If it leaks every time it rains then your chickens aren't going to benefit from the shelter and they will be more likely to suffer from health issues.

While your chickens could manage with a simple box, it needs to be a large enough box that they can live peacefully. If you make your coop too small for the amount of the chickens you are housing in it then they are going to get into lots of fights with each other and you shouldn't be surprised to find bloodstains on the coop floor. This is especially true during the winter when the birds are at their most agitated.

Speaking of winter, chickens don't like the cold. But they also don't like extreme heat in the summer. To balance this can be a little bit tricky but your best bet is to include a couple of windows that can be closed during the winter. The airflow will lower the temperature in the summer but when they are closed they will trap heat in, which makes them perfect for the winter. Just make sure that you put up some mesh wiring over the windows so the birds can't get out. But even more important than not getting out is keeping other animals from getting in.

One of the most important aspects of any chicken coop is its security. You know the expression "fox in the hen house?" I hope it simply remains a saying to you because to experience it in real life is to experience a massacre. Predators will kill your chickens if they are given the chance, so we need to keep our coop safe. Especially at night since this is when predators think they can get away with their devilish deeds. Keep your coop a foot off the ground with a ramp for the chickens to climb to get in. At night you must put your chickens into the coop and close the door. The door and the wire mesh over the windows should be enough to keep predators out but make sure you keep an eye on the structure to spot any new holes in the walls or floor before a predator does.

Chickens like to sleep at night in a spot of their own, what we call roosts. Different perches are built into the coop and these are then covered in hay or other soft materials for the birds to stay comfortable. Chickens appreciate having a space to themselves to sleep at night, which ties into their need for enough space to stretch their feathers.

Don't forget that stretching their feathers isn't simply something they do in their coop. Your coop should be situated within a fenced-in area. Chickens are much smaller than cows or even goats, so keep this in mind when making the fence. You might need to run some of that wire mesh over the fence to keep the birds from slipping out. Some people prefer to let their chickens be free-ranged, which means they aren't kept in a fenced-in location, but this can lead to problems with predators. In fact, it doesn't even necessarily need to be a predator that kills the bird. My parents' old dog once tore apart a neighbor's free-range bird before anyone could stop her. She wasn't trying to be harmful, it really did seem like she wanted to play with the bird, but the result was a dead chicken nonetheless.

Hitting all of these features is remarkably easy. A couple windows, a fence and a door that closes give you a chicken coop. If you only have one chicken then it could be one square foot and still work. But before you settle on a coop or start planning one out, you should take a moment and figure out how many birds you are planning to raise. As with most things, it is better to go with a bigger coop than you'll need when first starting out. It will give your birds more than enough space while affording you the option to increase the number of birds you're tending to as you get more comfortable with raising them.

How to Tend to Chickens

As mentioned before, chickens are easy to tend for. Your chicken coop will handle the most difficult part of their needs. Shelter and space and protection from predators are all handled by the coop, so what is left?

Well, those chickens need to be able to come out of the coop so you have to open up the door in the morning. This is a good time to check on their food and water and to see if there are any noticeable issues such as bloodstains. If there are then they could mean that a predator got in but chances are your chickens were fighting due to lack of space. However, the odds are that you don't discover any such thing and you can go about with the rest of your day.

The next time you tend to your chickens should be at night. Gather up your chickens and put them back into the coop. I prefer to check for fresh eggs at this point but others like checking in the morning. Either time is fine, there really isn't right or wrong here. Again, check their food and water. Assuming that both are fine, this makes up your daily routine of tending to your birds.

If they don't have enough water then you need to pour some more for them. Chickens tend to drink somewhere around a cup worth of water a day. They like to sip away at it slowly throughout the day. If you have six birds then a gallon of water should last about two days. Make sure you check the water level in the morning and at night, as too little water can slow down egg production as their bodies won't have the resources necessary to create their eggs.

Chickens also need food. Go figure, right? As mentioned, half a dozen chickens will eat about $30 worth of food over the course of a month. Simply check their food containers in the morning and at night. Top them up when necessary. Just like chickens need water to produce eggs, they also need food to produce them. Make sure you feed them the right food. Egg-laying chickens will need a different feed from meat chickens. Packages should be clearly labeled and once you have found a brand that works you can stick with it, so it's not very hard to figure out what you should be serving. If there is a problem with the food then fights between your chickens and signs of weight loss will make it clear.

If you can manage these tasks then you will be able to keep your chickens alive but there is one more step. Chickens poop a lot so you should go into the coop once a week to shovel it out and clean up the place. Check the nesting boxes to ensure they're still in fine condition with plenty of bedding, remove the manure and look for any signs of wear and tear in the structure of the coop. Plan to do this once a week, say every Saturday or Sunday. This will be the most time-consuming part of tending for your chickens but it is far from difficult. These animals are remarkably easy to care for and make a great introduction to livestock for beginners to backyard homesteading.

Chapter Summary

- Raising chickens offer three major benefits: meat, eggs, manure.

- Chickens are one of the more inexpensive livestock animals to start raising.

- Chickens can be grown for their meat. These chickens should be sold early, as keeping them alive exposes them to a lot of medical issues that would be cruel.

- Eggs are the best reason to raise your own chickens. A chicken lays roughly five eggs a week. During the winter they stop laying but they can be tricked into continuing with the right lights.

- Chickens poop a lot and while it is gross to clean, it can be used to improve the soil of your crops and your compost mixtures.

- If you want to raise chickens then you need a proper chicken coop.

- A chicken coop needs to be large enough for the birds to feel comfortable, otherwise, it will lead to fighting. It must also be secure so that predators can't get at them. You also want to have a window in place that can be opened and closed to improve the temperature.

- Your chicken coop should also open up into a fenced-in area so that the chickens can stretch their legs during the day.

- Check your chicken coop once or twice a day for eggs. Some like to check in the morning, others check at night and some check at both.

- You need to put your chickens into the coop at night for their own safety.

- Check water levels and feed levels, keep both well stocked.

- Chickens need to have the poop cleaned from their coops once a week.

In the next chapter you will learn what it takes to raise your own goats. Of all the animals in this book, goats are the most versatile and they can easily be the most profitable if you approach them with an open mind. You'll learn the many things goats help us to achieve, as well as how to tend to them.

CHAPTER SIX

RAISING GOATS

When's the last time you looked at a goat? I guarantee you that it's been too long. Take a moment and Google them. Look at how cute those little guys are. Isn't that a great time? Out of all the livestock animals we'll be looking at goats are definitely the cutest of them. They're fascinating creatures, too. Goats are actually as loyal as dogs are, which makes them a great pet alongside the productive qualities that make them wonderful livestock animals.

In chapter four we looked at how to convert a shed into a barn with a particular focus on the needs of goats. This was due to the smaller size of most sheds. They are often too cramped to make for good cattle barns but goats don't need nearly as much space. However, since we covered that topic we won't need to go into it in length here. A quick refresher will be adequate. Goats need shelter from the elements and bedding to sleep at night. They don't necessarily need to be put into the barn at night the way that chickens do but this can reduce the risks of predators getting at them. Of course, a goat doesn't have nearly as many predators as chickens do but the risk is still worth noting.

Goats are a more expensive animal than chickens. They need a larger barn and this typically costs more but beyond that they range in price from $100 to $300 each. While they do benefit from being fed hay and grain feeds, goats have a surprisingly versatile diet which actually makes for one of the major benefits. As you'll soon learn, goats have a remarkable array of benefits they offer backyard homesteaders. They have so many benefits, in fact, that I am comfortable with making the claim that goats are the best livestock animal for inclusion as part of your homestead.

In this chapter, we will look at these many and varied benefits to see just how wide-ranging they are. I'm positive that there will be several benefits which you wouldn't have guessed in a thousand years. With this taking up the first half of the chapter, the second half will move into how to tend to these productive and impressive animals.

The Benefits of Raising Goats

While each of the following is a clear benefit, it is important to note that you shouldn't expect to make use of each one. In fact, chances are quite good that some of these benefits might even be unappealing to you. Don't think that raising goats means making use of each of these. Figure out those that are most attractive to you and focus your attention on them. This will help to prevent you from trying to stretch yourself (and your goats) too thin.

Meat: Goat meat tends to sell for around a buck and a half, which is quite a bit less when compared to chickens. But unlike chickens, goats can grow to be quite heavy. A healthy goat grown for meat will often be around seventy pounds which means you can make close to (or more than if you're lucky) $100 each.

However, this is a hard tactic to recommend. Yes, goat meat can make you a decent amount of money and it is even quite delicious. If you haven't ever had it then I do recommend it. But the problem here is that goats are much easier to bond with compared to chickens and condemning them to the slaughterhouse feels an awful lot like sending your dog off to be processed. Many people who purchase goats with the intention to sell them for meat find that they are unable to do so when the time comes. However, goats are quite versatile creatures and there are several other ways in which they can earn you money or produce goods for you to use around your homestead.

Milk: Goats can be raised for dairy production. Despite being so much smaller than a dairy cow, goats actually produce a lot of milk. So much, in fact, that the chances are good that you won't be able to use it all. Goat's milk can be used to make just as many dishes and products as cow's milk, products like cheese and yogurt, creams and butter.

Goat's milk can be sold directly to customers but your local area should have codes and regulations in place that could make this difficult. Many find that selling the milk can be a bit of a hassle but it is much easier to sell the products that the milk helped you to make. For example, goat's milk chocolates have very few rules and regulations in place yet they sell like hot-cakes because of how delectable they are. Plus, there's one goat's milk product that is so impressive that it warrants its own discussion.

Soap: Goat's milk can be used to make soap and this is one of the most widely profitable products you can make on your backyard homestead. Goat's milk soap has gained popularity in recent years because it offers a more humane alternative compared to products that use animal fats. Soap made with goat's milk also uses fewer chemicals during the production process and this appeals to people looking for more natural products. On top of all of that, goat's milk soap is less irritating on sensitive skin and this is really its biggest selling point.

You can easily make your own soap to sell from the milk your goats produce. While you should have no problem finding local stores to stock your product, nothing says you have to. In fact, the most profitable approach to selling goat's milk soap (and other beauty products) is the large market for them online. It's easy to sell your products on sites like Amazon and Etsy. So long as you maintain a focus on producing high-quality products then you can be sure that you'll build up a customer base. I would love to say "in no time" but even high-quality products require time and dedication to get them off the ground. But, with that said, goat's milk products are wickedly profitable and I highly recommend looking into how to make and sell them yourself.

Fiber: Goats are furry little critters with rich coats of hair that can be collected. Some goats produce mohair while others produce cashmere. Both of these fibers are used in knitting and crochet. While mohair is more likely to be used in arts and crafts, cashmere has been used in clothing from some of the biggest brand names in the fashion industry.

The cool thing about collecting fiber from goats is how many options it allows you, the farmer, to participate in. If you want you can collect the fiber and sell it raw and in bulk. This will fetch a fair price depending on your local market. But if you have a creative bone in your body then this is a great way to get your hands on the raw materials you need to make anything from hats to pillows or dolls to stockings. With a bit of dye, to color the fiber, and some creativity, you can make arts and crafts that you can sell locally or online. There has been a recent surge in the number of people purchasing crafts online, fuelled by sites like Etsy, and a unique style will help you to stand out from the crowd. This approach is far more complicated than we'll be covering in this book (plus it doesn't particularly fall into the category of "backyard homesteading") but it can be extremely profitable, as well as extremely rewarding.

Clearing Land: Here's one of the most surprising ways to make use of your goats: Make them get a job. But not just any job; after all, it'd be a little weird to see a goat serving you breakfast at the McDonald's drive-through. Instead, there are two jobs which goats are best for. Clearing land is the first and it is shockingly profitable. The only issue is that it requires either a large amount of time or a large number of goats. The fewer goats you have, the less you can charge per acre cleared.

Goats naturally chew on everything. Grass, flowers, hay, even clothes and things you don't want them to chew on. It's just what they do, they're mean, munching machines. Well, not so much mean. But they do love munching on things. This has actually created a small industry that functions by renting out goats. Farmers, like yourself, will rent out their goats to landscapers or other farmers who are looking to clear some land. Rather than use heavy machinery, this is an all-natural approach to clearing land.

It might sound silly but this is actually a pretty amazing thing to see happening. It represents just how deeply some people are embracing the idea of natural farming. Goats aren't going to burn fossil fuels you damage the land. They're gonna eat up all the greenery they can and then pass it back out through their body. For those that are using them to clear farmland, they're both clearing the land and fertilizing the soil to increase the amount of nutrients present. It's a win-win situation for them. But for you, it's pure profit. Loan your goats out and make bank based on how many acres they're clearing. Large companies will rent out upwards of 250 goats and charge $700 an acre. Considering that the average farm has roughly 450 acres, even clearing one-tenth of that would earn back amazing money.

Carrying Gear: The other job that goats are great at is helping to carry heavy gear. While many people think of donkeys as pack animals, it is surprising how few people consider goats to be. This is surprising because it is not just hyperbole to say that goats make for the best pack animals around. Their only weakness is that they can't carry as much as a camel. Camels are the best if you are looking to carry heavy gear that you can't break up between multiple animals but goats remain the best in general for three key reasons.

The first is that they can carry around 30% of their body weight in gear. This can be upwards of forty pounds. That's a lot of weight off your shoulders if you're going on a long hike or journey. Of course, if you are going on a long expedition then you might consider this to be too light. However, you can always use more than one goat.

When you start increasing the number of goats you have with you, you might think this in turn increases the amount of weight you need to set aside for their food. But remember that goats will eat anything. They love munching on their greens. While having some food is always a good idea, you don't need to bring any with you. Even with no grains, your goats will remain healthy and they'll find plenty to eat along the way.

You might also think that adding more goats makes it harder to keep them together but trained goats are extremely loyal to their humans. Remember that these creatures are as loving and loyal as dogs. They can be trained almost as well, too. Trained goats know to stay near each other and their leader, so it is actually extremely easy to keep them together.

Goats are poor picks if you're looking to carry gear across a desert but if you're looking at fields, forest or hills then they're the best. Hills especially, as these little guys are extremely nimble.

Fuel: While most people won't find themselves drawn to this particular benefit, it is still worth mentioning. Goat feces can be burnt as fuel for fires. It doesn't have the best smell, so your neighbors might not appreciate it but goat feces certainly makes it easier to get a bonfire going.

This particular benefit is one that is more important to those who live in poverty or third world nations. However, it is worth knowing that this remains an option. In the aftermath of natural disasters, your goats may just prove to be the fuel that lets you stay warm throughout the night.

Hides: If you are slaughtering goats for their meat then you should make use of their skin as well. Remove the skin and let it dry out. You can tan it to get a texture close to leather like you'd get from cattle. You can use this in place of leather. Anything you could make with leather, you can make with tanned goat's skin.

Goat hide is also used across the world for various items. By leaving the hair attached, you can make a fine rug out of goatskin. In Africa they use the skin of the goat, along with the hair, to make heads for their drums. This isn't the most productive way to make use of your goats, nobody would raise them only to make use of their hides. But if you are going to be killing them for their meat anyway then it is best to make use of as many parts as you can so that you aren't wasting a life for naught.

Eating Scraps: Another reason you don't need to worry about bringing food for your pack goats is their voracious diets. Goats will pretty much eat anything and this includes the scraps from your table. You might not think this is a benefit but it helps you to reduce the amount of food waste you are making

Some food waste can (and should) be used to make compost. As a backyard homesteader, compost is necessary to create rich fertilizers for your plants. But there are lots of foods that shouldn't be composted in this manner. When this is the case, or if you simply have no need for more compost, you can use your table scraps to feed your goats. This is especially great for keeping down your budget (as it can replace the need to purchase goat feed), plus it makes it easier for feeding goats on the trail.

Manure: Much like chickens, goats make manure. Since manure is simply animal dung, you could argue that every single species of animal (man included) produces manure. But the important difference to make note of is the quality of that manure. It is common knowledge that chickens make wonderful manure that is high in nitrogen but did you know that goat manure is as well?

Along with the nitrogen, goat manure also has lots of potassium in it. Nitrogen and potassium are two of the macronutrients required for plants to thrive (with the other being phosphorus). Goat manure also has trace elements of many of the micronutrients and minerals which plants require, though they require far less of them when compared to the macronutrients.

A single goat will produce a bit more than a ton of manure in the course of a year, which should be more than enough for your plants. But one of the benefits of goat manure is that it is easier to collect. Chickens, for example, both poop and urinate at the same time. While their feces should hold its shape, it is much weaker and, speaking honestly, pretty gross. Goats, on the other hand, poop out little pellets and they can pee separately rather than having to at the same time. These are much more solid and easier to collect and it makes for a far less nauseating experience. (However, all animal manure loses its nauseating quality once you get used to handling it).

How to Raise Goats

As you've now seen, there are tons of reasons to raise goats. So the next important question is just how difficult is it? It is certainly harder than chickens but it isn't nearly as hard or expensive as cattle. Mostly you need to prepare your property in a manner that ensures they won't accidentally poison themselves when you aren't looking. It's this annoying habit they have.

We covered how to convert a shed into a goat barn, so we won't go over that in-depth here but it is extremely important. You can get by without providing a secure indoor shelter for them but this would be extremely cruel to the animals and it would increase the chances of predators making off with them. That shelter should be part of a fenced area that gives them enough space to roam around. Within this area, they will also need a separate location for eating. It is best to keep their food away from their beds. They enjoy having discrete locations for each of their activities and mixing places together is prone to agitating your goats and increasing their aggression. If you are milking your goats, which I highly recommend, it is best to have a separate area for this as well. Inside of your barn is best, as you can get some rather expensive equipment to help out with this and it is good to be able to lock it up at night. The inside of the barn should be packed with plenty of hay both for the goats to be able to sleep and as a midnight snack for when they get hungry. This area is kept clean and separated from the rest so the goats will feel comfortable and they can return to when they are feeling stressed. A healthy, relaxed goat produces far better than a stressed-out goat ever could.

There is a lot of equipment that people don't realize they need for goats, or for any other livestock for that matter. Things as simple as contains and troughs for them to eat from. To keep goats producing at their best you will want to get them a mineral feeder so that they get plenty of the nutrients they need to get their meat and milk fresh. While goats on long journeys can be kept alive with little food, those that are used for producing need to be kept healthy and well-fed. There are brushes to wash their fur, containers to collect the milk in, you may even want to take them for a walk like you do a dog. This can actually be a fun and important step in tending to your goats because it keeps them healthy and fit.

When preparing the pasture they are going to be living in, you absolutely must Google each and every type of plant you encounter. Some plants are poisonous to goats yet they will munch away on them all the same. They just love eating whatever they can get their teeth on. If it's a plant, they'll chomp away at it. Make sure to remove any plants that prove to be poisonous, otherwise, you are going to have a large bill with your local veterinarian and some pretty unhappy goats.

Goats can get sick for plenty of other reasons. They might be exposed to the cold for too long or kept out in the rain and develop a sneeze. They might have something happy internally that you've never ever heard of. Check your goats every day and watch their behavior. Look at their eyes and ears for signs of discoloration or leaking fluids. The signs that your goats are sick shouldn't be hard to miss so long as you pay attention to them and keep an eye out. When you spot them, don't try to fix them yourself. Seek a professional's help. They'll know what to do and be able to pinpoint the problem and provide you with a plan of action and chart a road to recovery.

Chapter Summary

- Beyond being adorable, goats are one of the most profitable and enjoyable livestock animals to raise.

- You will need a barn for them, more about which is found in chapter four.

- Goats have many, many benefits associated with them.

- Goat meat is tasty and it can sell for a pretty penny; goat milk can be used to make dairy snacks, drank as is or used to create soaps and beauty products for sale locally and online.

- Goats are great for providing fiber, clearing land as a natural lawnmower, carrying around heavy gear and creating fuel to make a fire in an emergency.

- Goats can be skinned for their hair and hides, they will eat pretty much any scraps you have left over and they produce a lot of manure which is rich in nitrogen and potassium.

- To raise goats you need a barn for them. You also need a field for them to wander in.

- Make sure to check all of the plants in your pasture to make sure that none of them are poisonous to goats. They will eat them regardless so it is up to you to prevent it.

- Goats should be given a place to sleep, a place to play, a place to eat and a place for milking. These should be kept separate from each other.

In the next chapter, you will learn the benefits that come from raising your own cattle. These animals are much more expensive to raise than any of the others in this book, so I don't recommend them for beginners. However, knowing what raising cattle entails is always useful for expanding your backyard homestead in the future.

CHAPTER SEVEN

RAISING COWS

Cattle are an expensive animal to raise. It easily costs you more than fifteen thousand dollars to set up cattle. Of course, how much you spend is determined by how many you plan to raise but with the higher cost associated with them, it serves you best to aim for several animals considering the cost. It is much harder to recommend that beginners start with cattle, but there are a few reasons that make them a particularly attractive investment to some.

We'll be looking at the benefits of raising these animals before we turn over to looking at how to prepare our homestead for them, and how we prepare them for our homestead. This will be a shorter discussion than goats or chickens, simply because a lot of this is far beyond the reach of beginners. Consider this a primer on why you should start slow and ramp up your production later. But if you are lucky and happen to be gifted with a cow or two or simply decided to start small and keep it that way, it's good to have the basics down.

The Benefits of Raising Cattle

Cattle are the most expensive livestock to start raising and so the benefits of raising them need to be weighed against their purchase. Those benefits are far fewer in numbers when compared to goats. They're about even with chickens, though cattle are worth quite a bit more money than chickens are when all things are considered.

The average price for a pound of beef is between $2 to $6. This is going to vary depending on the local area, as well as different factors such as if you are selling free-range beef or veal. Each of these is going to be a little more complicated than a normal beef cow. This price might seem a little low but when you consider the fact that the average weight of a cow for slaughter is one thousand and two two hundred pounds then you can see how this quickly changes things. You're looking at easily $2000 a cow. This makes them a good livestock animal for selling for meat, but they also eat more and so raising individual cows for sale really isn't worth the money.

If you are looking to raise cattle to slaughter and consume yourself then raising one or two at a time can be a viable option. The meat from a single cow can last you quite a long time, especially if you have a good deep freeze. This can save you a ton of money on meat. But if this was the only reason to raise cattle then it might not be worth it. Often it is a good idea for backyard homesteaders to purchase a cow for dairy and then slaughter it for its meat later on. It won't provide nearly as much meat but it serves a much better purpose.

Cow's milk is our main source of milk these days and for good reason. They produce plenty of it. Selling milk is hard to do in most states but you can use milk to make other dairy products that have fewer regulations surrounding them. For example, selling milk chocolate is much easier to do and you can typically make a lot more money doing so because people are willing to spend a lot more money for sweets. You will also be able to save money by replacing milk in your grocery shopping, this is especially true thanks to the sheer amount of milk a cow produces. It is unlikely that your family will be able to use it all.

Cows produce a lot of manure and this can help you to keep your vegetable garden filled with healthy nutrients for your plants. This is common for all of the livestock animals we've looked at so far but cows, being the largest, produce the most.

One final way of benefiting from cattle is to keep a hearty bull around. Bulls can be used to mate with female cows to produce offspring. These offspring can be sold to earn money but you can actually sell the service of the bull, too. People rent out their bulls as studs to impregnate other farmers' cows. A couple of hours of work can earn you a pretty penny this way.

How to Raise Cattle

The first step to raising cattle is to purchase some that are already healthy. Your biggest goal when raising cattle is going to be to keep them healthy because healthy cows are productive cows. This begins by first purchasing those that have a clean bill of health. Never purchase cattle without first inspecting them. You want to make sure that they are alert and that they show a reaction to encountering you. Take a look at their eyes and ears for any substances that are leaking. Get as close to them as you can and listen to them breathe. Signs of sick cows include wheezing, sneezing and coughing so if there is anything wrong with the cow's breathing then avoid purchasing it. If the seller can provide you with the documentation for any of their vet visits, this is perfect. Those sellers who are willing to share an honest medical history with you are people that you can trust.

Cows need a fair deal of space. Within their fenced enclosure, they will need an area for gazing. They love eating away at the grass so provide them with plenty. The fence will stop them from running away but it wouldn't actually be very helpful in stopping predators. If the creature can kill a cow then it can get over a fence. A barn will provide shelter from the snow and the rain. A cattle barn will have to be fairly large, as they are big creatures and they take up a lot of space themselves. But cows don't need to be put into the barn at night. Give them the choice to return if they want but you shouldn't be too worried about leaving them out.

When you purchase a cow, you need to then transport it. This will require you to get a trailer for hauling cattle. This experience is a scary experience for the cows and really stresses them out. Whenever you are moving them into or out of the trailer you should try to be as quiet as you can. This will help them to stay calm and not stress them out anymore than necessary. If you have cattle already and are purchasing more then it is a good idea to keep the new cows separated from the rest for a few days or even a couple of weeks. This will give you a chance to see if there are any noticeable problems with the new cows before they risk spreading it to the rest of the herd.

While grazing provides cows with a little bit of food, it isn't enough to sustain them and keep them healthy. You are going to need to have plenty of dry feed to keep them fed. And, trust me, cows eat a lot. You'd think with all the grass they graze on that they wouldn't need to eat so much but that grass is actually to help them be able to digest their food. It doesn't provide them with nutrients, it just helps them to be able to pass their food as manure later. A feeding area should be placed away from the sleeping area. Make sure that there is lots of water available. One of the ways that you can keep your cows in the best health is to fortify their drinking water with vitamins and minerals. This can be an easier way to improve their diet rather than experimenting with changing their diet. But with that said, a fortified feed can also be a smart idea.

Cattle get dirty often and I always recommend giving them a solid brushing on a daily basis. This helps to keep them clean but it isn't their hygiene that is so important. It is important, absolutely, but getting this close to your cows daily is the more important thing. This lets you bond with the cows over time so that it is easier to interact with them in the long run. It also lets you inspect them daily to see if there are any signs of sickness or injury. This can help you to catch issues early when they are still preventable. If you see any sign that your cattle are sick, such as those you looked for when first purchasing them, you should take their temperature and listen to their heart rate and their breathing. This is information a vet will ask you about, so it is always good to be able to provide them with as much information as possible. Another thing to watch when you can is how much they are eating. One way that you can spot issues quickly is to notice changes in the animal's diet.

That's raising cattle in a nutshell. This is a topic where there is still a thousand times more information to learn but it is also the hardest of the animals to raise. If you are new to farming, start with chickens or goats. Turn a profit or feed yourself with these animals first and then consider if you even need to invest in cattle farming. With its difficulty and cost, you might find that it just simply isn't worth it.

Chapter Summary

- Raising cattle is extremely expensive and it is incredibly hard to recommend it to beginners.

- Cows offer the same types of benefits that chickens do. They can be raised for their meat, for the consumables they produce and for their manure. Chickens lay eggs, cows provide milk.

- If you are looking to breed your own cows then keeping a healthy stud around will be necessary. If your stud is of high-quality then you can make some money loaning it out to other farmers for breeding purposes.

- Only purchase healthy cattle. If you suspect anything is wrong with the cattle you are considering, or if you get the impression that the seller isn't being fully honest with you, then pass on those cows.

- Cows need a lot of space, so a big pasture with plenty of space for grazing is important.

- You must provide your cattle with a barn to protect them from the elements.

- Cows need to be transported with a cattle trailer. When introducing a cow to your farm make sure that you are extremely quiet. Hooting and hollering at them will only stress them out more.

- Isolate new cattle from your herd for a month or two to ensure they are in full health. This helps

BACKYARD HOMESTEAD

to prevent any illnesses from passing into your main herd.

- Cattle get dirty quickly and can use a quick clean every day. This isn't necessary but I recommend it so you can get in close with your cows, bond and inspect their health on a regular basis.

In the next chapter, you will learn the many benefits that come with raising your own honey bees. From healthy honey to a hand with your crops, these creatures are amazing. You'll also learn how to raise them yourself so that you can benefit from them and help to keep them from going extinct.

CHAPTER EIGHT

RAISING HONEY BEES

Honey bees might be a surprising animal to think of livestock but they can be an incredibly valuable part of your backyard homestead. They provide you with honey, no surprise considering their names, and this can add some sweetness to your life. It can also be sold for a decent amount of money and the wax can be collected for use in products like beard balms or other health products.

Just like in the last two chapters, we will start out by looking at the benefits of raising honey bees. This will have one or two surprising points in it that I doubt you had known about. There are more ways to earn money from your honey bees than you might have thought. From there we will move into a guide on how to raise them. They aren't a particularly hard animal to take care of so long as you give them enough space and set them up properly to avoid the elements.

The Benefits of Raising Honey Bees

Honey bees are small animals that don't take up a lot of space. This alone is a major plus about raising them. They also don't take a lot of money. Getting started with honey bees will require you to buy a certain amount of equipment such as homes for the hive, a smoker and a beekeeper's suit. These start-up expenses will cost you a decent penny but ultimately honey bees are cheaper to get started with than even chickens. This cheap cost will be balanced by their space needs, which we will discuss later in the chapter.

For now, let us content ourselves with looking at why you would want to raise honey bees in the first place.

Honey is Healthy: I don't need to tell you that honey is sweet and tasty, I'm sure you already know that fact. But did you know that honey is ridiculously healthy? Instead of adding processed sugar to your coffee, try adding some honey. It has such a long list of nutritious ingredients that it requires a paragraph of its own!

The nutrients include calcium, copper, iron, magnesium, manganese, niacin, pantothenic acid, phosphorus, potassium, riboflavin and zinc. Considering that honey is sweet enough to stand in for sugar, using it is literally trading out a substance that wants to kill you for one that wants to ensure you stay healthy. And there are a ton of health benefits associated with all of these. Studies have shown that honey can boost memory recall. It has been used in helping coughs and even in treating wounds. There is research pointing towards the idea that honey can help to increase our white blood cell counts when undergoing chemotherapy, though it is safe to say more research must be done in that department. Honey is also good at keeping you energized and it can even be used to kill bacteria and reduce dandruff.

Considering most of us just think of it as tasty, it sure is an impressive substance.

Honey Makes Money: Plus that's just a great rhyme. You can collect, bottle and sell your honey. Local businesses are a popular customer for bottled honey but you can sell it directly to customers yourself. A gallon of honey will make you around $50. However, honey must be strained to have the wax removed from it before bottling. This will reduce the weight of the honey you pull from your bees but it, too, is profitable.

Bee's wax can be used to create lip balms or beard balms. These can be sold to stores or individuals but the biggest market for their sale is online. Bee's wax also finds use in eye shadow, blushes, sand salves, moisturizers, creams, pomades, as well as lots and lots of other beauty products. Bee's wax is a natural product and if you keep your creations all-natural then they can sell for a higher price and appeal to a larger demographic. People love buying natural products. The time for disgusting chemical concoctions has passed.

Honeycomb can also be sold as a snack. It is a delicacy that some restaurants will serve with meats and cheeses. You can always sell it directly to customers but restaurants are a more profitable source of income in this regard. Though, of course, it should go without saying that you will want to try this snack for yourself as well.

Honey Doesn't Spoil: Mushrooms go bad eventually. Vegetables go bad in time. Fruit goes bad. We can freeze them to increase their lifespan but this only lasts so long until time takes its toll and decomposition sets in. This is an inevitable part of life. Try as we might to avoid it, we simply can't.

But honey doesn't spoil if it has been processed. Unprocessed honey that still has its imperfections will last about a year. But if you process your honey to remove these imperfections then it could outlast you, your children and your children's children. Processed honey is pretty much the longest-lasting resource you can produce on your backyard homestead.

Pollinating Your Fields: Depending on what you are growing, you may need to pollinate your plants yourself. This is more common with fruits than with vegetables, as they tend to produce small flowers or buds which need to be pollinated to start the fruiting process. There are self-pollinating varieties of most fruits but a lot of them require you to use a brush to carefully pollinate the buds. This can be a time-consuming process and it is always easy to miss a few buds. But your bees have your back.

Bees naturally pollinate flowers. They go from flower to flower, soaking up the pollen there to help them in producing their honey. As they do this, they naturally take the pollen from one flower to the next, thus pollinating them for you. This makes it much easier to ensure that your fruits will be coming in fully this season.

Renting Them Out: This directly builds on the previous benefit. Because bees are pollinating machines, they are extremely useful to farmers. If you have more bees than you need to pollinate the crops on your backyard homestead then you can always rent them out to farmers to help with their fields. This particular approach doesn't earn a whole lot of money but you keep all of the honey they make and basically just make like $50-$100 a month for keeping your bees on someone else's property.

It's a pretty win-win situation all around.

They're Low Maintenance: Bees are extremely easy to raise. You barely need to do anything. Setting up a hive takes a bit of work but even then, the bees are the ones that will be doing the most of it. Once they are up and running then you'll want to check on the hives once a week or so to make sure that it is staying clean and that there are no signs of infection or warfare (between other insect species).

When it comes time to collect the honey, this will take a little more of your attention. Maybe a couple of hours or so. Say half a day if you are extremely distracted and don't yet know what you are doing. But this only happens about twice a year. Assuming that it takes you six hours to collect the honey, you're looking at about 36 hours of maintenance in the course of a year. Considering that a year has almost 9000 hours, that's not too shabby!

Bee Conservation is Extremely Important: Bees are quickly becoming extinct. This is a horrifying thought because they are responsible for so many flowers getting pollinated. Without bees, the world will be a much less colorful place. I hope I don't need to explain to you why losing many of our flowers would suck. All that natural beauty…just…gone.

It's distressing to consider.

But raising our own honey bees is one of the ways we can help to solve the problem! These creatures do tons and tons of work for us and they ask for very little in return. They want a little house to home them but that's it. They like to be set away from people. They don't like to be bothered but they also don't like to bother people either. They are very friendly, which is not a trait they share with hornets or wasps. But bees have no desire to hurt you. They just want to help spread flowers and pollen throughout the world.

If any of these benefits seem appealing to you then I want to strongly suggest that you raise your own honey bees. Doing so not only provides you with lots of healthy honey (plus all those other benefits) but it helps to ensure a species doesn't go extinct and that the world keeps some of its color. That alone makes raising honey bees a truly rewarding experience.

How to Raise Honey Bees

Farming isn't just about animals you can raise for meat anymore. More and more mini-farmers and backyard homesteaders are realizing that raising honey bees is simple, compliments their other farming ventures and can make them a lot of money. But, just like with the chickens, you are going to need to check your local zoning laws to see if you are legally allowed to raise your own bees. Some places have laws against them; as silly as these are, we need to respect them. Begin by first confirming you can raise them on your property before continuing.

If you are legally allowed to raise your own honey bees then you are going to need to pick out a space for them. While your goats, chickens and cows just need a large, mostly flat space, your bees are going to be more specific in their needs. Picking a space for your honey bees is more like picking a space to grow a crop. You'll want them to have full sun for part of the day but they'll want to have shade for the second half. Bees also need lots of water, however, you can't just provide this by adding a drip feeder like you would give a hamster. It is best to position your beehives close to freshwater whenever possible. If you don't place them near natural water then you'll want to set aside a bucket or two close to them. When you skip this step you'll find that there are a lot more bees hanging out around your other animals' food area and this can stress out both animals.

Along with water, you are going to want to position the hives so that they aren't going to be assaulted by heavy winds. One of the ways this is done is by housing the hives inside a wooden structure. If you've ever seen a beekeeper pulling honey out of a wooden box then you've seen these structures. The bees don't necessarily need to have this structure but they will thank you for it. If you don't house them this way then they are going to be exposed to the wind, rain, snow and other predators. For their own safety it is best to provide them with this. Plus it will also reduce the chances that they end up taking off for a better home. When positioning this, you will want to keep it fifty feet away from the rest of your homestead if possible. This is that privacy that they enjoy. Avoid placing them near any areas that see a lot of foot traffic. If you don't have the space to separate them at this distance then you can use fences or hedges to help reduce the amount they are disturbed. This can be a smart approach in general, too, as it helps to keep them away from your guests.

Bees should be purchased and installed in the spring. We purchase bees not by the individual but by what is called a frame. These are then slotted into our bee boxes. They can easily be slid out to check on the health of the bees. The frames are filled with honeycomb. You can purchase bees that come packaged with a caged queen. This takes a while to get the colony up and running but it is among the cheapest ways to begin beekeeping. You can also purchase a nuc, which is an already established but young hive that has anywhere from two to five frames. This is more expensive but it can be the easiest starting point if you aren't patient enough to wait for a caged queen to get pumping out the babies and establish the colony. A nuc will come with a queen that has only just started to give birth, which is always a good thing. When purchasing bees, always purchase from a reputed seller. I recommend starting with a nuc. You could dive in even deeper and begin with a full colony but I don't recommend this for beginners. It is better to start with a nuc as they have the possibility of having a honey harvest-ready before the end of the summer. This isn't always the case but it would be at least a year before you could expect the same starting smaller. Let your nuc grow and expand and then add more colonies once you understand and have a sense of the struggles that come with beekeeping.

Installing bees in your homestead isn't a one-size-fits-all kind of operation. Each breeder will have their own recommended steps based on what they are selling you. This means the size of your purchase, how many bees there are, how old they are, all of these factors will play into how to best set up your colony so always listen to what the seller says. While you're preparing to get them in, make sure that you purchase a smoker and a beekeeper's outfit. You might be able to get by without purchasing a smoker right away but I would recommend that you do. However, you absolutely must purchase the beekeeper's outfit. This one can't just be purchased later. It will make your life a thousand times better to deal with bees with the proper equipment.

While bees will mostly take care of themselves, you do need to give them some help when you first set them up. When you install a rack into your bee box, this isn't the same as transporting the bees' entire home. It's more like you cut their bedroom out of their home and moved that. There are a lot of repairs they need to make to make this new box into a home. When they get in, the first thing they are going to want to do is to start sealing up any cracks in the box. This is done to protect the queen and prevent any intruders from making their way into the box. However, this takes a lot of food and energy to achieve. Bees survive on nectar but they aren't going to be getting nearly enough for what they want to do. Not if you want them to finish their repairs quickly. We get around this by making our own nectar for them. Grab a jar and fill it up to the halfway point with sugar. Pour water into the jar until it is full and then give it a stir. This will create a mostly sludge-like mixture. Purchase a feeder lid and attach it to the bottle. Place the jar in the bee box upside down so that gravity forces it all down. Your bees are going to drink through about a jar of this a day. This is a lot of sugar you'll be going through but it is only temporary. It should take about a month for your bees to finish their job. As they get closer to the end, they will be drinking less and less of this mixture. Eventually, in around a month, you will want to remove the jar. Bees that are fed this way produce a less tasty honey, so it is best to encourage them to head out and find nectar themselves.

Check your hive once a week. This should only take about thirty minutes. You want to see how it is building up its structure. You'll want to see if the queen is laying eggs properly. Feces is also going to build with time and excessive amounts can cause problems. Cleaning it away is as easy as scraping it out. While you're keeping, look for signs of struggle such as dead insects and bees. Research foulbrood and make sure the queen's eggs don't look anything like it. As the bees continue to grow, they will start to need more and more frames. A bee box should only have about eight frames in it. If you are looking at more than you are going to want to get another bee box to support the new growth. Keep in mind that a smaller hive has more difficulties keeping themselves safe, so it can be a smart idea to wait until you can move a fair-sized colony into the new box.

You might be able to get a honey harvest in that first summer but if you can't then you will have one in the following spring. It really depends on how many bees there are and how well they adapt to their new home. You can purchase specialized gear to extract honey but this doesn't have to be a high-tech process. A clean scraper can be used to cut the honey from the frames to catch into a container at your feet. This collects the honey and the wax at the same time. You'll need to separate these two with some cheesecloth. The honey should be then prepared while the comb can be stored or used for other products. Let all the bubbles rise out of your honey before bottling it.

And there you have it. From preparation to harvest, all you need to take care of honey bees. I hope you'll consider them.

Chapter Summary

- Honey bees have so many benefits that they are second only to goats.

- Honey is ridiculously healthy for you and packed with so many nutrients it is almost insane. Processed honey also pretty much never goes bad.

- Honey can earn a lot of money and you can even sell the honeycomb as a delicious snack.

- Bees are amazing because they pollinate your fields for you. Rather than requiring you to purchase self-pollinating plants or pollinating them by hand, your bees will take care of it for you. Because of this you can actually rent your bees out to farmers to help them with pollination.

- Bees take very little effort to care for but doing so helps to preserve their numbers. This is important as bees are having an incredibly hard time and are facing extinction.

- Purchase or make a bee box. This should be placed about fifty feet away from any of the other major sections of your homestead. If you can't give them this space then use walls or hedges to give them some privacy. This will also cut down on the wind.

- Bees should be kept near water. If there is no water to keep them near then provide them with a couple buckets of water next to the hives to keep them from having to go looking for their own and landing in your livestock's drinking water.

- For those beginning, purchase a smoker, a beekeeper's outfit and a nuc. A nuc is a small colony of bees that has already been started. This will make for the easiest starting experience.

- Bees need lots of energy to make their hive when they are first installed. Take a jar and mix it half full of sugar then fill it with water. Attach a feeder lid to it and keep it upside down. Keep this full at all times for a month. Once your bees stop making their hive and start reaching out they won't need this anymore. Remove it, as sugar-fed honey tastes worse than honey produced from nectar.

- Check your hive weekly for signs of trouble. Clean away any excess feces on the bottom.

- Honey is harvested about twice a year. Separate the honeycomb from the honey with a piece of cheesecloth, then sell or use both products.

FINAL WORDS

From why you would want to start your own backyard homestead through to planting your crops and tending your livestock, we've covered everything that goes into this. But a book like this can only take you so far. I could tell you exactly how to grow a tomato, down to a day-by-day calendar of what to expect but it wouldn't be any more useful to you. The issue here is that you need to get your hands dirty and do this yourself.

Of course, to call that a problem is a lie. It isn't a problem. It's an absolute delight. Growing your own food, watching as your livestock grow and develop, all of it is an immensely rewarding experience. There is a sense of pride that starts to develop when you take your first steps in this direction. With every additional step you take, this pride grows and grows. There is a reason that farmers and those that work with nature in this way are so content. It can be hard work at times, especially when you're shoving animal manure or dealing with a pest infestation, but that hard work feels earned. It's not the same as the exhaustion that comes from working an office job or cleaning dishes all day. It's an exhaustion that feels right, that is fixed with a deep sleep at night so you can get back at it tomorrow.

BACKYARD HOMESTEAD

I don't want to suggest that you run out and quit your day job. This would be a silly response. As with anything, it is best to ensure you have a steady income in place while you are getting started. But as your backyard homestead grows, so too will the money that you can make from it. In time, you can make enough to support yourself and your family. It's always possible to purchase more land if you need to but this comes in time. Master your own backyard first, then start to branch out.

Remember that this isn't a race and there is no finish line. This is a way of living that you can take with you until your last day. It can reconnect you to the world around you and give you a much richer experience. Many people find that as they get older they feel as if they have lost something, some connection to the world and nature. And I'm not talking about getting to be sixty or seventy. I'm talking about hitting thirty and forty. We have much deeper connections to the world when we are young but then the responsibilities of school and work seem to suck that away. Many people find they need something in their lives to reconnect them and this has led to a resurgence in people running their own backyard homesteads.

Maybe you feel like you are missing this connection. Or maybe you simply want to save some money on groceries or earn some back each year with the harvest. Whatever reason brought you here is the right one. It is your personal reason and so long as it speaks to you then it is valid. Take the lessons that you learned in this book and start to put them into action. Take your time and learn as you go, building up each new piece of knowledge one step at a time. Before long you'll be running and you'll have an amazing backyard homestead to show for it.

I hope that your harvests are always plentiful and your animals healthy.

MINI FARMING

A Beginner's Guide to Profiting from Crops, Vegetables and Livestock

By Luke Smith

INTRODUCTION

There is a misconception in popular opinion that farming refers to large scale operations while gardening is the light version of farming. Gardening, in this sense, is seen as almost entirely providing food for the gardener, often at the expense of a money and time investment. The home gardener is able to supplement their diet and save money on vegetables that they would otherwise have bought, but it isn't considered a profitable venture. If you want to make a profit, then you need to get into farming, but that requires a lot of starting capital to purchase land for the crops, machinery, and labor to work them, as well as legal documentation. A gardener might be able to sell a couple of heads of lettuce at the farmer's market without a problem, but a large scale farming operation requires the farmer to jump through a lot of hoops.

One of the reasons this particular misconception is so potent is the economics behind it. If you try to set up a small farm by following the practices that the industry uses today, then you are going to find yourself breaking even at best. Unfortunately, the odds are stacked against this, and it is much more likely that you will lose money on the endeavor. The problem with turning a profit on a smaller venture in this way is the fact that the techniques used by the agricultural industry are designed for large scale operations. The economics of scale allows for much larger numbers during the production process since the yield of the harvest will be large enough to make up for them. With a small farm, this is just not possible.

But purchasing a large farm isn't viable for 99% of the population. We don't have the resources to invest in buying more land and setting up enough crops to turn a profit. So, it would seem, farming should be left to the professionals and us small-scale growers should stick to our little gardens.

But, thankfully, that isn't true. It is one perspective that seems to jump out from the information available to us, but it isn't the only one. To use a popular metaphor, this perspective sees the glass as half-empty. However, the problem isn't that small-scale farmers can't make a profit. The problem is that small-scale farmers can't make a profit using the techniques and methods of larger-scale farms. This glass-half-full perspective points us towards a new approach, one that *can* profit us: Mini-farming.

Mini-farming uses techniques to compress space and cost to allow farmers more control over their crops or livestock. It is a small-scale approach to farming, but one that is able to net a profit because it doesn't waste time or money on techniques that are outside of its budget. What's even cooler is the amount of variety that mini-farming offers the farmer. Yes, we might set up a crop in a fashion reminiscent of traditional farming, but we could also use hydroponics, raised garden beds, or even simply use containers for our plants. A mini-farm might be grown in your backyard, or it might be grown inside your house. The number of options available is quite high, and when we turn to these techniques, we find that the whole enterprise is much more profitable.

In this book, we're going to cover mini-farming from top to bottom so that you can get started with your own productive and profitable agricultural enterprise. In chapter one, we'll look at the benefits of mini-farming so you can assess whether it is right for you or not. Chapter two will move into crop farming to cover topics like monocropping, crop rotation, raised crops, hydroponic crops, and more. Chapter three continues this discussion by turning to the profitable vegetables that make up those crops. Chapter four will move away from crops to discuss specialty livestock such as goats, cattle, chickens, and bees. We'll have a single goal throughout all these conversations, which is to provide you with the knowledge you need to start a mini-farm that will earn some money. To that end, chapter five looks specifically at profits to see how that money is earned through selling. Finally, we'll close out with a discussion about preventing pests from ruining our farms, and how we can best maintain a high standard of excellence in our venture.

If you have ever wanted to take your gardening skills to

the next level and earn money with them, then what are you waiting for? Stop studying the large operations outside of your budget and start mini-farming to maximize your investment and make money.

CHAPTER ONE

WHY MINI-FARMING?

We've already touched briefly on several of the reasons why you might want to start using mini-farming techniques. In this chapter, we'll go into each of these (and more) in greater depth. It should be noted beforehand that these reasons don't exist in a vacuum, but rather they interact with and affect each other in a fluid way. This means that we should consider these reasons as building on each other to create a dialogue that encapsulates the subject as a whole. This approach is necessary, as we can't separate these reasons from each other when we put mini-farming into practice. In some ways, that could be seen as a negative. If there was a particular reason that you disagreed with, you can't exactly step away from it or avoid it. However, I'm sure you'll consider each of the following points as benefits rather than limitations.

Mini-Farming Requires Less Capital

Farming can be an extremely expensive undertaking. If you approach farming without a concern for the money involved, then it won't be long before you see how much more expensive it is than you imagined. Sure, you're just growing plants in the ground or taking care of livestock, but these create tons of expenses. Some are clear and easy to spot. Others are hidden, only to reveal themselves when you suddenly need to invest more money into the enterprise. This can absolutely devastate your bank account if you aren't careful. A consideration of the costs involved can help to illustrate this.

To begin with, you are going to need seeds. The larger the crop, the more seeds you'll need. Seeds need to be planted in soil, but not just any soil works; it needs to be healthy and nutrient-rich. The crops need to be tilled, and then planted. They'll need to be watered and fertilized. These are all costs you'll need to sink into the farm. You'll also need to pay for the land you are growing on. If you are raising livestock, then you need to purchase food and water for them. They'll need some shelter from the harsher weather conditions. They need to be looked after, kept in good health, and tended to on a daily basis. This can eat up time, or it can eat up money if you've hired an employee to help you. Small amounts of livestock typically don't cause many problems, but the more you are raising, the more likely you are to encounter zoning laws and other issues that need to be resolved beforehand. Of course, legal issues tend to cost a pretty penny themselves. Plus, we haven't even considered the effort or money it takes to keep everything clean and in working condition. And this isn't even bringing in hidden costs such as transportation or taxes.

It can be very, very expensive to start a farm. What's even worse is the fact that most people who are looking to start one are going to need to start small. Just as most businesses need to start small and take time to grow, so does a farm. The end goal might be to become the largest agricultural farm in your local area or even your country, but this takes time to build. When you're starting, if you approach the farm the way you would one of these bigger ones, then you are going to lose a lot of money.

But mini-farming techniques aim to reduce the amount of capital needed to get started. They focus on creating smaller, but more profitable farms. To approach a new farm from the perspective of a large-scale operation will only lead to frustration. These farms are already large enough to be able to afford the steep costs. Their size alone gives them confidence in their ability to turn a profit. But if you are just beginning, then you won't know how much money you'll make with each harvest. Instead of starting large and investing lots of initial capital into the venture, starting with a mini-farming approach will benefit you. You keep the costs low while you go through the process of bringing your first few crops to harvest and getting marketplace data to make profit projections with confidence.

Even if you are aiming to be the biggest, starting with a mini-farm is undoubtedly the way to go. You'll be able to make back your investment and can consider the experience as a hands-on education in the business of farming.

Mini-Farming Requires Less Space

When you look at those larger farms, it can be truly impressive the way they sometimes stretch on for miles at a time. Walking through all of the fields of one of these large-scale farms can easily take up half your day.

However, while there is certainly something very powerful about traversing a farm of this size, it can also be a depressing experience if you are looking to start your own. It would be wonderful if you could have a farm of that size straight out of the gate, but the cost would be colossal. If you always needed that much space in order to start a farm, then nobody ever would. Most of us understand that this is unrealistic to begin with, and so we start at a much smaller scale, but this, in and of itself, is not actually mini-farming.

If you've ever talked with new farmers about how they are beginning, you might have discovered a disappointing idea that seems to reoccur. A lot of new farmers think they need to replicate the planting practices of their colleagues who have larger areas. Yes, they have much less space, and so it is done at a smaller scale, but they still design their fields after the pattern of those larger ones. This isn't necessarily the worst idea in the world; you can still grow some lovely vegetable crops this way. But it is like trying to raise a horse inside your house. Sure, it might fit through the door, but you aren't really using the space available to the best of your ability. For our purposes, this approach is equivalent to traditional farming, but just at a reduced size. It captures one element of mini-farming (the mini part), but it still relies on techniques and approaches designed for larger farms. It is technically using less space, but it isn't using the space in a way that produces benefits from the smaller size. It doesn't turn a potential weakness into a

MINI-FARMING

strength.

Mini-farming can be done in various ways and in very different locations, both in size and in environment. Let's say you plant a crop in the backyard, directly into the ground. For simplicity's sake, we will pretend the crop's dimensions are one square foot. This crop takes up that full square foot, and nothing else can be done with this space. If you plant following traditional guidelines, you will only have two rows with two plants each, at best. This space could be used much better. If we tossed out the traditional approach, then we could use what it is called intensive gardening. In this method, plants are grown much closer together. This reduces the amount of space needed, and in doing so, it reduces the amount of care (watering, fertilizing) that needs to be done. We could implement this technique directly in the ground, but it benefits from using a raised garden bed, which brings up another way we can save space. One of the cool things about raised garden beds is that you can design them to have multiple levels. When grown in the ground directly, you have a one square foot crop. If you used a raised garden bed design that had multiple levels, then that same square foot could allow you to raise three times as many plants in the same space.

Reducing the amount of space your crops take up reduces the amount of money spent on soil and fertilizer. It allows you to more easily tend to your crops, as they're packed tightly together instead of spread out. It also allows you to make a higher income because you can earn more money per square foot. So, by using mini farming techniques, you not only reduce the amount of money that you are spending on your farm, but you are also increasing the amount it is earning you. This is an impressive double-whammy that makes mini-farming an attractive prospect.

Mini-Farming is More Productive

To say that mini-farming is more productive is not the same as saying it is more efficient. For that particular reason, you will need to keep reading. Productive here is used to mean that mini-farms produce a larger harvest than traditional farming does. This might seem a little confusing. After all, if it produces more, then why aren't these techniques being more widely used? The answer to this is surprisingly simple. Larger farms have been in operation for much longer, having started their business prior to the scientific revolution that is the mini-farming movement. The older the farm, the more likely it is that old-fashioned farming techniques are being used. Rather than fall into the trap of tradition, we can implement these techniques to achieve massive yields.

When it comes to the gardening component of mini-farming, there is plenty of research to show that intensive gardening practices achieves more abundant results. Likewise, there is also a lot of research that has been done on the use of raised bed gardens. Both of these methods have proven to be more effective than traditional approaches, and this improvement becomes exponential when you practice intensive gardening in raised beds.

This productivity comes at a cost, however. While the yields are bigger, intensive gardening has been shown to have a detrimental effect on the environment. Because of how closely spaced plants are, this approach requires more fertilizer per square foot. Now, to be clear, it requires less fertilizer overall because it is done on a smaller scale, but the amount of fertilizer in any one section is more. To return to our one square foot garden, we would need to fertilize the four plants in the traditional approach, but we could be fertilizing five to ten plants when grown intensively. The problem with this is that it allows for more fertilizer to run off and potentially damage the environment by getting into the water or causing the pH levels of the nearby soil to spike. Neither of these is very positive, and they may be enough to put some farmers off mini-farming.

But if you are careful in the way you design your farm, this won't be a problem. The biggest issue is the possible damage that comes at the cost of productivity, but we've already identified one way of reducing this. If we design raised garden beds that are at least a foot tall, then we can use a solid bottom to prevent too much fertilizer from getting out. We'll still want to have a drainage hole or two in the raised bed, but this won't see nearly as much run-off. This is achieved through the height of the bed, as the raised bed is entirely filled with soil. That means there is more space for fertilizer or water to have to move through before it exits back into the natural environment. That goes a long way to reducing the negative impact that mini-farming could have while keeping it productive as can be.

As a side note, keep in mind that mini-farming techniques can also be applied to tending livestock. Just like crops, mini-farming can reduce the amount of capital and space you need to tend to your livestock. However, it is important to be aware of the dangers of overcrowding your animals. Livestock raised through mini-farming methods aren't necessarily more productive than those raised traditionally, but if you overcrowd them, then you will see a reduction in productivity. Avoid this by being aware of how much space your livestock have. Remember, they are living animals with their own nutritional needs and health concerns. They need to be kept healthy to be productive. So while livestock may not particularly benefit from this one, they certainly do from the next one.

Mini-Farming is More Efficient

MINI-FARMING

Efficiency is all about achieving more by doing less. Does this mean there are fewer steps involved in mini-farming when compared to traditional farming? Nope, it is exactly the same in that regard. Of course, there will be variance from farmer to farmer. One person may not like to use any pesticides while another does. These aren't differences in the required steps. They are more like modifications. They are determined by the personal preferences of the individual farmer making the choice, and so, we can disregard the variance they introduce to the equation. But this raises an important question: If there is no notable difference in the steps taken between traditional farming and mini-farming, then just how precisely can mini-farming be described as more efficient?

The answer to that requires we take into account our role

in the maintenance and care of our farms. Pretend for a moment that you are about to grow a crop using the traditional approach. The first thing you are going to have to do is till the soil. So you go up and down and create rows. Then you need to seed the bed or the field. Assuming that you aren't using machinery, you are going to need to physically go up and down the field to sow seeds yourself. Of course, you're still going to need to sow seeds when using the mini-farming approach. The difference isn't in the action required there, but in the walking and travel time required to achieve the goal. You need to walk through the field or the bed. But mini-farming uses garden beds that are tightly packed and closer together. Depending on the size of a particular bed, you might not need to do any walking in order to sow the whole thing. While walking might seem like the least of your concerns, the time it eats up can become considerable by the time harvest rolls around. Then you need to go through the field again to harvest all your vegetables, far more time-consuming compared to having them close at hand through intensive gardening techniques.

Another aspect in which the efficiency becomes clear is through our use of raised garden beds as part of our mini-farming toolkit. Raised garden beds will take a little longer to set up since you usually need to build them yourself, but they greatly cut down on the total amount of maintenance and work to be done. A field needs to be tilled between each crop to keep the soil rich and healthy enough for the plants. But a raised garden bed doesn't need to be tilled. The soil is maintained through simple additions of compost before shutting down for the winter. When combined with liquid fertilizer applications, this ensures a nutrient-rich soil without the need to waste energy tilling. Plus, sowing seeds is much easier in a raised bed garden, especially when intensive gardening methods are utilized, because everything is self-contained and easy to access. So here again, mini-farming increases efficiency by cutting time-consuming corners.

This follows through to our livestock as well. If you are raising a bunch of cattle in a large farming operation, then they're either going to be in a tightly regulated barn or out grazing in the fields or the yard. Just like with the larger crops, if you want to check on each of your animals, then this takes up much more time because there is more space in which they can be spread out. But mini-farming reduces the total amount of space and keeps everything closer together. Again, it may seem like a minor detail. It is negligible if you are only cutting out one trip's worth of walking, but the increase in efficiency is best seen by considering the total amount of time saved. It adds up. You can easily save a dozen or more hours in a single season this way. If you continue to farm for several years, then this number can reach into the hundreds or thousands a lot quicker than you might expect.

Mini-Farming Gives You More Control

When it comes to gardening, the best control comes from hydroponics or indoor methods. Growers can fine-tune the environmental factors that have a positive or negative effect on the plants they are growing. This level of control may be present in a mini-farm, as hydroponics can make for a wonderful addition to a mini-farm. But comparing a traditional farm to a hydroponic setup would create a false equivalency. We could say that hydroponics is better than ground-grown farming, but we couldn't use hydroponics to argue for mini-farming as a whole. To do that, we need to compare similar methods. So no hydroponics and no raised beds. This leaves the question: Does mini-farming still give us more control when we use intensive gardening techniques directly in the ground?

The answer is a resounding yes. This harkens back to the second section we looked at in this chapter and the way that mini-farming uses less space. We know that less space allows us to be more efficient, but this isn't the same as control. But it functions on almost the exact idea. The reduction in space makes it easier for the farmer to get at and check each of their plants. By having everything more tightly grouped together, we can get in to check for signs of pests or disease and other issues without as much hassle. This will give us an early warning sign of issues so that we can act quickly and solve them. Thus, mini-farming gives us a higher level of control over the health of our crops. But that isn't all.

One of the more typical ways of maintaining a traditional farm is to use heavy equipment to work the fields. There is something captivating and enjoyable about riding in a tractor, watching how the equipment tills the field, or harvests the plants or fertilizes the soil. A lot of craftsmanship went into the design and creation of these tools. They are also invaluable when it comes to bigger crops because they reduce the amount of time necessary to complete any particular job by a significant amount. But what we save in time we lose in control. We aren't tilling, sowing, fertilizing, or harvesting the field. We're letting the machinery do that. Sure, we're behind the wheel and telling the tractor where to go, but it is the machinery that does the physical work involved. So by turning to this equipment, we hand over our control.

But we've already seen how mini-farming allows us to save time. Even if we weren't walking the larger field, but using a tractor, we would still find ourselves spending more time working the larger field compared to the smaller, mini-farm we've built. So we don't need to worry about time when it comes to mini-farming. That means we don't need to call in the heavy machinery and let it do the job for us. Mini-farming requires us to be directly involved in each of the steps leading to harvest. That allows us to learn more and improve our farming skills, while also giving us full control over the farm and what happens. A piece of heavy machinery such as a harvester might chip a blade and result in a jagged cut that ruins the vegetable, but we will have this experience when we practice mini-farming. If something damages our vegetables, it will be our own doing. We can learn from the mistake so that we don't make it again. Taking over more control means that we gain more experience farming while reducing the frequency of unforeseen problems arising from our reliance on technology.

Mini-Farming is Simple

The best part of mini-farming is that it doesn't take a ton of knowledge in order to get going. You are working with much smaller spaces, and this means there is much less you need to worry about. You are still going to need to know how to check the pH level of your soil or how to clean up after livestock. But the smaller size reduces the need for heavy machinery, and so you won't need to worry about having a tractor license or knowing how to use a combine harvester, for example.

The hardest part of mini-farming is cutting through all of the discussion you'll find on the internet or in most gardening books. If you look at how far apart to space a crop, most books are going to tell you a number in inches or feet that is based on traditional farming. This stands in contrast to intensive farming and mini-farming techniques, and if you go researching without this in mind, then you may be tricked into thinking that you aren't tending to your farm properly. Unfortunately, this is one piece of misinformation that can be quite frustrating. But if you're aware of this pitfall, you'll be prepared.

For the most part, you won't need to go out and seek a whole lot of knowledge. You'll want to learn about the crops or animals you are raising, but you don't need to worry about machinery or the logistics of harvesting a big field and preparing the yield for transportation or storage. These are elements that you'll have to tackle in a miniature form, which makes them far easier. In a way, you could consider mini-farming sort of like a set of training wheels for your farm. Only, you don't actually need to upgrade further down the road. Your mini-farm can be made to turn a constant profit without the next for expansion. So not only is it simpler to get started, but it can remain that simple throughout the rest of your time tending to your mini-farm.

MINI-FARMING

Chapter Summary

- There are a lot of reasons to use mini-farming techniques, most of which are interconnected with each other.

- Traditional farming is an expensive endeavor that only typically makes a profit when it is done on a very large scale. It requires the purchase of land, equipment, seeds, and labor.

- Mini-farming uses less space to grow more plants, doesn't require heavy machinery, and can be harvested by an individual. This all contributes to reducing the amount of capital you need to get started.

- Mini-farming also requires far less space than traditional farming does. Plants may be grown closer together using intensive gardening techniques, or the use of raised garden beds can allow for multiple levels of plants grown on the same spot, thus greatly increasing the number of plants that can be grown.

- Using less space will also lead to less money spent on fertilizer or soil. You will need to use more fertilizer per square foot, but there are less square feet in total, and so the amount will still come out lower, so long as you are growing with productive techniques like rotating crops.

- Traditional farming requires you to have access to the ground itself. A mini-farm can be run through a hydroponic system, and therefore, you could start one indoors if you had to.

- Mini-farms are more productive than traditional farming as it produces more in a smaller area, and this results in bigger yields come harvest time.

- Intensive farming practices can lead to excessive runoff of fertilizers, but combining intensive gardening techniques with sufficiently deep raised garden beds can compensate for this.

- Mini-farming doesn't just refer to crops, but livestock as well. However, livestock will produce far less if they aren't given enough room to live in. You can still raise livestock with mini-farming techniques, but you should be mindful of ensuring they have enough room to stay healthy.

- Mini-farming keeps everything tightly grouped together in a smaller location, which results in a reduction of time spent traveling from plant to plant to take care of them all.

- If you use raised garden beds in your mini-farm designs, then you also remove the need to till the soil, which improves efficiency.

- It's easier and quicker to tend to your livestock when they are kept close together through mini-farming techniques.

- Traditional farming techniques require heavy machinery, and they take away a level of control from the farmer. Mini-farming techniques offer this level of control back and make it easier for farmers to get a hands-on look at precisely how well each plant is growing.

- Mini-farming requires less equipment and therefore has a smaller learning curve, making it a great choice for beginners. The impressive result that mini-farming yields also makes it a terrific choice for experts, too.

In the next chapter, you will learn all about crop farming. We'll look at terms like monocropping, crop rotation, mixed cropping, intercropping, hydroponics, and raised bed gardens in order to see their various advantages and disadvantages. Often, the best mini-farms are those that make use of multiple approaches to balance these pros and cons. By knowing these, you'll be able to design the most effective mini-farm possible.

CHAPTER TWO

CROP FARMING

Despite the fact that we are discussing mini-farming, our available approaches to crop farming are anything but mini. We have just as many, if not more, options available to us when it comes to how we raise the crops in our mini-farm. This is just another reason why this is such a wonderful approach for beginners or experts alike.

In this chapter, we will look at six of the available options we have for growing crops. While we'll look at each of these individually, they are actually broken up into pairs. The first pair is monocropping and crop rotation, which make up two sides of the same coin and deal with how often a crop should be planted. The next pair is mixed cropping and intercropping. These two techniques are often discussed interchangeably, but they are two different approaches to mixing different crop types. The final pair is hydroponic and raised bed crops, which together represent two approaches to how we can go about raising a crop beyond just planting directly in the ground. Every crop is going to either be monocropped or rotated, though not every crop will check off the other four categories. However, a single crop could fit into up to three of these categories (selecting one from each pair). So, while these are all unique, they can be combined if you wish.

Monocropping

Monocropping is the act of continuously growing the same crop in the same field year after year. Of all the categories we look at, monocropping is the one that is the least recommended. That's because it has such a negative impact on the environment. In this regard, it stands in stark contrast to the practice of crop rotation. It still has its supporters, but we are seeing more people discussing how harmful this practice is, and critics of it have even begun to put pressure on major farmers to switch from monocropping to crop rotation. We'll start with the positives so we can see why this technique still has its supporters, but once we turn to the negative, it will be clear why you should never consider monocropping as an option for your mini-farm.

The biggest argument that people make for monocropping is that it is more profitable because it saves money. If you are going to monocrop, then you can really niche down and only buy the equipment necessary to look after that one type of crop. On a larger farm, this can mean an absolutely astounding amount of money saved because the heavy machinery that farms use is costly. Right out the gate, this results in a lower initial investment, and so it is easier to make a profit. Since you plant the same crop every year, you also save money on a recurring basis. You only need to purchase one kind of seed, and you already have the necessary equipment, so the only additional costs for further crops are the seeds, labor, and any mechanical repairs that are needed. This makes monocropping attractive to those who are primarily concerned about earning money. But monocropping has a problem which is far too big to ignore.

That problem comes in the form of the environmental impact that monocropping has. When you grow a single crop in the same soil year after year, the nutrients of that soil are depleted. Eventually, if you tried to grow a crop in this soil, it would just starve to death because it doesn't have the nutrients for healthy growth. Yet this is exactly what monocropping does, and so it becomes necessary for the farmer to use chemical fertilizers to feed the plants. When we talk about issues like fertilizer later in chapter six, we'll be looking at organic fertilizer. Organic fertilizer is weaker than chemical fertilizer because it isn't replacing the nutrients in the soil; instead, it's supplementing them. But monocropping drains away all those soil-bound nutrients, and so it requires a strong fertilizer for the crop to come to harvest. The problem with fertilizers of this strength is that they actually do more damage to the soil and so they only make the problem worse. It's a little like drinking coffee late at night. You might get the energy you need to keep going for a little bit longer, but you are going to crash hard if you don't keep drinking more. Instead of drinking coffee, it is better to sleep. When it comes to our crops, it's better to give the soil a rest through the use of crop rotation. But before we turn our attention to rotating our crops, there's another downside that can't be ignored.

Monocropping exposes your crops to a higher degree of risk of infestation by pests. Most backyard gardeners or indoor gardeners are at risk of pest infestation, but the risk is lessened due to the fact they often use a variety of plants. Not every pest wants to munch on every plant. I might like cake, but you might like pie, despite the fact we're both human. We see the same eating variances within a singular kind of pest as well. Perhaps this is no more clear than when we look at the aphid. Annoying little bugs that they are, aphids actually come in a bunch of different species. Most are named after what they prefer to eat the most, and so we have potato aphids, melon aphids, cabbage aphids and bean aphids to only name a few. If you are monocropping, then you are creating an immense buffet of their favorite food for these pests. In contrast, if you use mixed cropping or intercropping, then you create smaller pockets of the same plant, and this makes it harder for a particular pest to spread through it all.

Likewise, if you practice crop rotation, you will also reduce the frequency of pests. Monocropping plants in the same crop year after year mean your field can become a haven for pests. They can set up shop nearby, so they never have to travel far for a meal. Therefore, you have infestations year after year, and it can be a major headache. To solve this problem, you are going to need to use pesticides, and these contribute further to the negative environmental impact that monocropping has. Rather than just excessive chemical fertilizer in the soil, you end up with excessive amounts of pesticides (which may still be organic but shouldn't be present in such high volumes).

Ultimately, monocropping might save money in the initial investment, but it comes at too high an environmental cost to be recommended. The issues with pests can create a recurring problem that greatly affects your experience and will cut into your profits. If we compare the positive with the negative and then consider the mini-farming approach, we see that for our purposes, monocropping is too much of a negative experience. We don't need to purchase expensive equipment when mini-farming and so we wouldn't be saving money by monocropping. All we would get out of it is low-quality soil and, probably, a lengthy fight against infestation, so we might as well cross out monocropping and push it from our minds.

Crop Rotation

Monocropping, by growing the same crop in the same place, year after year, results in environmental issues and increased struggles with pests. It can also possibly lead to more soil-based disease problems. If we want to avoid these, then we need to make sure we rotate our crops rather than repeat them. Crop rotation requires you to grow more types of plants than monocropping, typically at least three or four kinds. While we will consider crop rotation using a traditional field planting approach, keep in mind that this technique easily adapts to raised beds or other mini-farming approaches to growing.

Before we can begin to rotate crops, we need to divide up our field into three or four sections. Often this is done, not within a single field, but across several, so that each field is growing a different kind of vegetable. If we go with four different sections, then we might have lettuce growing in the first, potatoes in the second, carrots in the third, and celery in the fourth. Already this offers a distinct advantage over monocropping through the variety it lets us grow. However, if we grew the same vegetables in each section year after year, then all we have done is monocropped four plants rather than one. What we need to do is rotate the crops. So, if they were in the above order during year one, they would then shift the second year so that the first crop was celery, followed by lettuce, potatoes, and carrots in the second, third, and fourth sections respectively. This would repeat each year until the crops were back to their original space.

Rotating your crops in this way is beneficial because it reduces the frequency of infestation since the parasites aren't given the chance to become accustomed to one particular type of plant in that location. This also has a beneficial effect on the quality of the soil, as different plants soak up different amounts of nutrients. One crop might eat up lots of nitrogen while another prefers to snack on potassium. However, most farmers want to be extremely specific about what they grow in each plot so as to avoid planting members of the same family next to each other. When this happens, there is very little benefit in crop rotation as the same nutrients, pests, and diseases are going to be affected by the change.

When researching crop rotation, you will often find it recommended that you plant a legume, a root veggie, a leafy green, and a fruit-bearing plant in the four crops. This is intended to keep the crops varied, and prevent them falling into the trap we just explored. However, it isn't an ideal guideline to go by because it isn't detailed enough. That can be seen by looking at carrots and parsley. Carrot is a root vegetable, parsley falls into the category of leafy green, so according to the commonly advised method, a crop of parsley could follow after a crop of carrots. But both carrots and parsley are members of the umbelifera family, and so they soak up similar nutrients while also attracting the same diseases and pests. When dividing up your crops for the purposes of rotating, go by the plant family rather than any loose descriptor such as leafy green or root vegetable.

The biggest downside to crop rotation is that it is more time-consuming than monocropping, and it can also cost much more. This cost is more readily apparent when considering larger farms since mini-farming doesn't require heavy machinery. However, you need to plan out your crop rotations ahead of time and learn how different plant species grow and what needs they have as far as fertilizer, watering, sunlight, temperature, and humidity go. So, while it is healthier to rotate your crops, it is much more involved than monocropping. With that said, I strongly recommend that you rotate your crops as this will keep your soil healthier longer and result in less headaches trying to fight off pests or diseases.

Mixed Cropping

Mixed cropping brings us to the second section of the chapter. Both mixed cropping and intercropping can lend themselves to monocropping, though they also benefit enough from rotating your crops that I am going to categorically advise you to stick to crop rotation in your planning. This section of the chapter should help to dispel some misconceptions around mixed cropping and intercropping since these two methods are often used interchangeably, despite the fact that they are quite a bit different. But they both have the same core idea, which has led to the confusion around them. Regardless of whether you are using mixed cropping or intercropping, each is a way for you to plant two crops in the same field rather than taking two fields to grow the same crop. Of the two, mixed cropping is the most chaotic because the plants being grown share pretty much the exact same space rather than simply sharing the same field.

This method was originally undertaken as a way for farmers to increase the likelihood of their crops making it to harvest. When you are growing food for personal consumption or for business purposes, having a crop fail is the worst thing that could happen. When you are growing to eat, having a crop fail means you might starve. When you are growing for income, having a crop fail means that you aren't going to make the money you need for your bills, or even to support your family and sow the fields the following year. Where we might be out a little money, but can learn from our experience, a failure of this sort could literally be life or death to a farmer. In order to reduce the risk of a crop failing, mixed cropping was used to double the chances of a successful crop making it to harvest. While the same land would be used, it is planted with at least two crops, which means doubling the chances. Some farmers will even go up to three crops in the same field, though it is important to select the right plants. Before we turn to picking our plants, let's take a moment to explore the difference between mixed cropping and intercropping.

The best way to understand what we mean by mixed cropping is to pretend we are about to sow a field with seeds. Normally we would go through the process of tilling the field and getting it into rows. Then, we would take our seeds and either plant them directly in the soil or scatter them about; the method we'd choose would depend on the requirements of the seeds we decide to plant. When we practice mixed cropping, we do the exact same thing. Only this time, we take two types of seeds and sow them at the same time. In a traditional field, this would result in rows that are growing two different types of plants in them. Of course, we aren't necessarily going to be seeding our mini-farms in this same practice, but we can use intercropping the same way. When we grow using a raised bed, we don't need to bother tilling the soil, but we can still spread out two different types of seeds in the same fashion as described here. It is just easier to understand mixed cropping in a traditional sense, as it highlights most clearly how odd this practice seems. Why exactly would mixing your crops together help to create a better harvest?

For that, we need to consider the many reasons a crop fails. Poor temperatures, too much water, too little nutrients, all of these are just some of the reasons. But if each plant prefers different nutrients and has different water requirements, what kills off one plant might actually be beneficial to the other. The result of this is a better chance that one of the crops makes it through to harvest, even if the other crop fails. Unfortunately, this has a bit of a downside to it too. Because these plants are going to be competing for resources, they aren't going to grow as big or bountiful as they would if the crop was only one kind of vegetable. The smaller yield can make this technique undesirable for those who are only growing to earn money, but those who are growing for food can help to prevent themselves from starving this way. But if you do want to give mixed cropping a shot, then it is important that you carefully pick which crops you are going to use.

Picking the right crop isn't very hard, though you do need to take in some considerations to determine what makes for a good pairing. To begin with, you will want to select plants that grow at different heights. Have one that grows taller than the other so that both plants don't need to fight each other for the sunlight. The taller crop is going to get more direct sunlight while the short crop will use the taller crop for shade, so make sure that your shorter crop doesn't require very much direct light. With that said, if you grow a tall crop, then you might want to grow a crop that climbs such as tomatoes, cucumbers, or eggplant, as they can use the taller plant for support. If you are growing three crops in the same field, then pick one that is tall, one that is short, and one in the middle. By doing that, you've created a natural tiering system that allows each plant to get enough sunlight.

You will also want to consider the way the roots of each plant grow. If you are cultivating a plant with a short root system such as lettuce, you wouldn't want to pair it with another plant with a short root system. All that would do is crowd the soil in that particular area and encourage plants to fight each other for resources. It's better to select a plant that grows nice, deep roots to go with one that has shallow roots. While you are considering the needs of the root, also consider how much water each plant needs. If you pick a plant that needs a lot of water, you should pair it with one that doesn't need much at all. What this does is protect you from the two extremes of the elements. Heavy rain might damage the less thirsty plant, but it will keep the thirsty one happy. But the thirsty plant won't be able to survive a drought while the less thirsty one might. The idea of pairing plants based on their needs continues through to their nutrients. If one crop wants nitrogen-rich soil, pair it with one that doesn't need as much nitrogen. By considering the needs of the plants, you can pair them in such a way that they increase their chances of getting to harvest while also not competing too strongly with each other.

If you have picked your crops well, you will see quite a few benefits. For one, if both survive to harvest, then you will have a variety that you wouldn't have otherwise. Of course, one of those crops might fail, but that doesn't mean that the field as a whole has failed. You've created a sort of safety net which would otherwise be absent. Depending on what you decide to grow, this method can actually help to keep your soil richer and for a longer period. Crops that are leguminous, for example, help to keep the nitrogen levels in the soil at healthy levels so you could be improving the soil rather than just draining it. Mixed crops have a tendency to attract less weeds because the way that the plants create a canopy helps to reduce the amount of sunlight that reaches the ground. That means less of this vital resource is being spent helping weeds to grow, and this means less time you spend having to weed the field. Likewise, there are less pests to deal with when you employ mixed cropping because there aren't as many of the same plant grouped together. Pests will prefer to snack on one of the plants rather than both of them, and the confusion that mixed cropping creates in the parasites helps to slow their spread and keep them out of the field in general. Finally, while you are less likely to see an increase in yield, it isn't unheard of. If you have a good year and the crops were chosen in such a manner as to not compete with each other, you may find that you are harvesting more than you expected at the end of the year.

With that said, mixed cropping more often results in a smaller yield. That's one of the reasons people may choose not to use the practice. Mixed cropping also takes a lot more research in order to pull off because you need to not only worry about the needs of one plant, but rather, you must consider the way that two different crops will interact with each other. This makes it a better technique for those who enjoy researching their crops and giving careful consideration to a handful of variables. It is also a lot harder to harvest a mixed crop because you need to be extremely careful. You could damage one of the crops while harvesting the other, and this would destroy the benefits. But if you are looking to maximize the space you have and increase the variety of what you are growing, then mixed cropping might be the right choice for you. But as it is intended for minimizing the dangers of crop failure, if you are looking to increase your yield and potential earnings, then you are better off going with intercropping.

Intercropping

Intercropping is quite similar to mixed cropping except that it separates the space between the different crops you are growing much more clearly. With mixed cropping, we saw that we cultivate seeds from both plants in the same row. Intercropping takes a little more time and dedication to get right. We would start intercropping the same way as we do a regular field or mixed cropping. Once we have our rows lined out, we would then sow the seeds of our first plant in the first row. We'd skip the second row and move to the third row and sow our seeds again. Once we have gone through the whole field this way, we would turn around and go back through it, hitting the even-numbered rows that we originally skipped. This time we would sow the seeds for the second vegetable that we are growing in the field. This would create a pattern of crop 1 - crop 2 - crop 1 - crop 2.

That is what we call row intercropping, as it requires the

rows in the first place. But it isn't the only form of intercropping, just the easiest way to grasp the concept. One thing that we could do is to create rows, and then sow the seeds of our main crop in each row. Then we would go back and sow the seeds for our second crop in the middle between the two rows. Yet another way that we could use intercropping is called relay intercropping. Relay intercropping is very specific about when the crops are planted, as the goal is to create a harvest period that lasts much longer than if they were planted at once. Finally, there is also a type of intercropping called mixed intercropping. This approach is a little bit more like mixed cropping, but instead of rows, you have different and varied patterns created from how you plant. Mixed cropping sees the two plants grown together, but mixed intercropping sees the plants grown close together in a way that is beneficial for growth. That might include planting a low and shade-loving plant next to a larger plant that could create a canopy to offer protection from the sun.

This separation between the two plants is the big difference that separates mixed cropping from intercropping. We still need to be extremely careful in considering what plants we are going to grow through this method. While they aren't using the exact same piece of land, they are grown so close together that you still need to take into account a lot of variables. Both plants grown in this method can have shallow or deep roots, as they won't be competing with each other in this fashion, but you will still want to keep them of different sizes in order for each to get enough sun. Most important of all is the way the plants use the nutrients in the soil. If you can balance them so that they aren't soaking up and using all of the same nutrients, then you will have a much easier time bringing the plants to harvest. It should be clear that intercropping uses up more space than mixed cropping, but this space can still be scaled in a similar fashion. When seeding a raised bed, for example, just be careful to sow seeds in straight line patterns so that intercropping is achieved rather than mixed cropping. If you find a few plants have strayed out of their area, then these can be trimmed away to promote better uniformity.

Intercropping takes longer than mixed cropping because you need to be more careful during the sowing stage, but it comes with enough benefits to make it worth it. Your crops will have less problems with weeds. That's because the taller plants will offer shade and make it harder for weeds to get sunlight. Also, the soil will be so busy with so many plants there won't be as much room for weeds to edge in. Intercropping can have more issues with pests than mixed cropping does, but this technique results in fewer pests than regular crops. Pests that prefer the first crop are going to have a hard time navigating through the second crop to get to their next meal. Disease is also less of a problem, especially disease that is found in the soil. Now that we've mentioned the soil, note that it will be able to maintain a much stronger structure due to the different types of organic matter on top. The structure of the soil is even further assisted if you decide to mix crops with different root lengths. This is beneficial for the soil, but it isn't a necessary step the way it is with mixed cropping. Since you balance the nutritional needs of the plants, you can make sure your crops aren't draining the fields. Again, this is made even more effective when you plant a crop like legumes that feed nitrogen back into the soil. Finally, intercropping results in a larger yield than mixed cropping tends to, and this makes it the better choice for those looking to earn money from their crop. So with all of these benefits, it is likely that intercropping is the way you'll want to go if you choose to grow in the soil. Let's take a look at the steps we take to intercrop successfully and then turn our attention over to hydroponics, a water-based approach to growing, and raised bed gardens.

Picking the right plants for intercropping is a lot like picking the right plants for mixed cropping; only there are some important differences that can make or break the whole endeavor. When we use mixed cropping, it is to ensure that one of our plants survives. Since this is the primary goal, we use plants that have quite different needs from each other, so that one of them will be able to survive, no matter if the season is full of heavy rain or a heat-wave drought. But intercropping is used to increase our yield in general. To achieve this, we need our plants to be closer together in their needs, so that we can more readily attend to them and make sure that they are getting what they need. That does leave us open to losing the crop if we plant thirsty plants and then have a drought, or if we plant drought-resistant plants and then have heavy rain. So the risk with intercropping is higher, but it is more in tune with what we would expect out of a traditionally planted crop, and so, while the risk is more than mixed cropping, it is still within the normal range expected by a farmer.

Start by researching the needs of your plants. This is easy enough to do; simply open up Google and search the name of your plant followed by "planting guide," "intercropping," or "nutritional needs." You'll get hundreds of sites with the information you want. First, look at the family that the plant is from. You want to avoid growing plants from the same family. Next, look to see how much water they need. You want them to need roughly the same amount, as you will be watering them together rather than separately. As it can be beneficial to grow plants with root systems of differing lengths, try to get one that is long and one that is short. Of course, you can get away with growing plants of the same size root system because they aren't directly taking up the same space they do with mixed cropping. You need to consider how the size of the plants will affect the sunlight. If you're planting tall plants, you may want to grow them in horizontal rows rather than vertical ones so that they don't throw shade on the short plants. But then again, if you pick shorter plants that prefer the shade, this might be precisely what you want. You may also want to consider planting species that have different growing periods, as having plants that are in different phases can help to reduce how quickly nutrients in the soil are used. Finally, consider making one of the species a legume or another type of plant (such as green manures or accumulators), which helps to keep the soil rich and healthy. If you follow these steps in picking your plants, you can see a major yield.

Harvesting is going to be easier with intercropping compared to mixed cropping, as the plants are spaced out better, and you don't need to worry about damaging one kind while harvesting the other. If you are mindful of your plants and look after them throughout the season, then intercropping can be one of the most effective ways to maximize the use of your space so that your mini-farm can thrive. We've been mentioning fields, but intercropping can be done in a raised bed, which we'll look at now as we turn our attention over to the last section of the chapter.

Raised Garden Beds

A vital component of mini-farming is the maximization of space. This results most often from the fact that we don't have as much space for farming as the major companies do. That pushes us to use practices like intercropping or mixed cropping, as these make better use of the space we have and allow us to grow two (or more) crops in the same area. But before we even begin to start sowing seeds, we should take the time to consider that space and see if we are using it to the best of our ability. One of the most effective approaches that we can use to achieve this is to adopt raised beds into our planning. Out of all of the techniques we've discussed in this chapter, raised beds are the most useful and highly recommended. In fact, frankly, the only way that we can get away with monocropping is through the use of a raised bed. In this section, we'll see why they are so highly recommended, by comparing them to traditional ground-based farming, as well as how we can quickly make one, and the many benefits they provide. But before we even get into them, let me say that raised garden beds should be the cornerstone of your approach to mini-farming.

But what exactly is a raised garden bed? This particular question might seem irrelevant, as the name serves as a description of the product itself. But a raised garden bed is essentially a garden bed that has been raised up above the ground so as to give the gardener more control over the growing environment. In a lot of ways, a raised garden bed is almost like a big plant container. We take a material like wood or stone, and we build four walls, being mindful of the size of the bed. We then fill the raised bed with healthy soil and plant our crop. If done correctly, this gives us a much greater level of control over the growing environment, and this makes it easier to keep the soil filled with nutrients. We can use these raised beds to raise monocrops, intercrops, or mixed crops.

The reason that we can actually achieve a healthy monocrop with a raised bed is directly tied to that level of control. Typically, when growing in a field, you are relying, to a considerable degree, on the soil that is there. We might add some fertilizer to the soil and give it a till or two between seasons, but this only does so much. If we are growing deep-rooted vegetables, then what happens, is the roots grow first through the healthier soil, which we added to the field, until they push into the natural soil that is beneath our added layer. In this approach, we give up control over the soil beyond our top layer. But, in a raised garden bed, we have the option to rely more fully on our soil. We do this by making the raised bed much taller. A raised bed should never be less than half a foot tall, and often, it benefits us to go a foot or even a foot and a half. Since we are the ones that fill it with healthy soil, this lets us control a much greater amount of the growing environment.

Before we look at all of the benefits, let us take a minute to consider how we build a raised bed garden. There are two approaches that differ only slightly, but it is important to consider which approach is more effective. Every raised bed will have four walls around it. These are typically made of wood, but some people may use rocks, bricks, or other materials, including some as wacky as old tires. Since we are growing vegetables to sell for human consumption, we need to be especially careful when it comes to picking a material. Tires, for example, can leak harmful chemicals into the soil, which then degrades the quality of the soil (and the nearby earth), and this can make vegetables grown in a raised bed of this nature pretty unhealthy. So, we must first pick a healthy material such as a hardwood or a non-treated wood. Next, we have to make the biggest decision: do we want to build a bottom for the bed or not? If we only make raised beds that are half a foot or a foot off the ground, then we should probably not add a bottom. That's because we want to make sure our vegetables have enough space for their roots to grow unimpeded. But, by choosing not to add a bottom, we find ourselves reducing the level of control we have over the growing environment because we need to rely on the natural soil underneath it.

I believe it is best to make raised beds with a bottom. The most obvious reason for this is the fact that it lets us keep our control over the growing environment. If we have a raised bed that is tall enough, there is enough room for the roots of any vegetable to grow, and this means that we don't need to cross our fingers and hope the soil is good. We'll see in a moment how we control the soil and keep it nice and healthy enough to achieve a monocrop. But, the other reason for including a bottom, is that it helps reduce the incidence of burrowing pests getting into our crop. While approaches like intercropping and mixed cropping can reduce pests such as aphids or whiteflies, they can't completely protect us from critters. A raised bed with a bottom does, as it becomes much harder for them to get in. It should be noted that, while it is harder for them, it's not impossible. That's because the inclusion of a bottom does not entirely cut off the raised bed from the natural soil. We need to be careful to ensure that our raised beds have enough drainage capability so as not to trap water and moisture in the bed. Trapped moisture makes it more likely that we'll drown our crop and lead to diseases such as root rot. We need to drill in a few drainage holes in the raised bed. We use a mesh covering over these holes to make it harder for things to get in, but the fact that there are holes at all makes it a possibility, albeit a rare one.

The biggest downside this approach has is that we need to build our raised garden beds. It requires a little bit of investment, as we need to acquire materials, as well as the knowledge necessary to put the bed together. It is also necessary we keep in mind that the larger the bed, the more securely it needs to be built as the soil we fill it with puts pressure on the foundation, and it could easily break one that is made cheaply or poorly. This investment of money, time, and energy may leave you thinking that you're better off skipping over raised garden beds, but that would be a mistake. Making the raised bed may take up a lot of time, but once it is made, it is far easier to maintain crops in a raised bed garden, as well as the soil itself. When you are considering using a raised bed approach for your mini-farm, remember not to design the beds to be too wide. We want to be able to reach all of our plants, and so we should never make raised beds that are wider than four feet. They can be as long as you want, but the width should be capped at four feet maximum.

So, there is only one major downside to using raised beds in your mini-farm, but what about the possible benefits?

These far outweigh the negatives, so much so that it isn't even fair to compare the two. It would be like comparing Shaquille O'Neal to someone who just picked up a basketball for the first time in their life. These benefits include the level of control we have, specifically over the soil, a reduction in pests and critters, less strain on our backs, fewer weeds, more space, and an earlier start to the growing season. If you are looking to maximize your potential profits, then that last point is going to be especially interesting to you. We'll take a look at each of these now, and then turn our attention over to the chapter's final section on hydroponics.

Let's start with that control. To begin with, we choose the soil that fills our raised beds. This allows us to ensure that we are picking soil with plenty of nutrients. At the end of the growing season, after we have harvested our plants, we need to prep our raised beds to ready them for the winter. While the easiest way to do this would be to cover them with a tarp, we can repair the soil by covering the beds with a layer of compost followed by a layer of mulch. The compost will decompose over the course of the winter so that, by the time the new season comes around, the nutrients have been able to permeate the soil and keep it as healthy as it was when we first used it. It is through this act of winter preparation that we keep our soil fertile enough to be able to monocrop if we so choose.

Because the soil is raised and kept separate from the

natural soil below, we reduce the incidence of weeds, pests, and disease. We may still find pests in our raised beds, but they have a much harder time getting there, and we can even use a ring of cinnamon around the edge of the bed to make it that much harder for pests to get in. Diseases like root rot or nutrient burn are still pretty standard in raised beds, but this is because they primarily happen due to the way the farmer has acted rather than just occurring naturally. Root rot happens when we overwater our plants, and nutrient burn happens when we feed them too much fertilizer. But, if we take diseases caused by our actions out of the picture, raised beds have far, far fewer problems with disease than traditional crops do. They also don't cause as many problems with weeds. When growing a crop normally, we need to till the soil of the field before planting each year. Doing this pushes around weed seeds and assists them in spreading out over our fields. Raised beds don't require any tilling, and so we don't move these seeds around as much. We also use a mulch covering during the winter, which helps to prevent weed seeds from getting into the bed, and this also blocks those that are present from getting enough sun and water to grow. Fewer pests and weeds, and less disease, means you don't need to devote as much time to saving your crop. You can have more time to literally enjoy the fruits of your labor.

Since we can design our raised beds in such a way as to create multi-level tiers, we can make the most out of the space we have. Even when we don't create multi-leveled setups, we can plant our crops closer together in a raised bed when compared to the same crop directly in the earth. That's fantastic because we want to make sure that our mini-farms are as productive as possible. But the greatest benefit of using raised beds in this fashion isn't one that affects your crops, but rather one that affects you. Since mini-farming doesn't use the same equipment as a larger farm, harvesting tends to mean getting down on your hands and knees to get to the same level as the plants. If you have bad knees or a bad back, then this can be the most miserable part of the whole experience. But a raised bed reduces the amount of strain that you need to put on yourself. With that said, this does affect certain vegetables such as potatoes. When harvesting potatoes, it is easy to damage them and thus reduce your overall profits. But raised beds make it easier to control the harvest and reduce damage such as this.

Speaking of profits, if you are farming for income, then a raised bed may help you to earn more. Because it is raised off the ground, these garden beds begin to warm up earlier in the year when compared to a traditional field. That means you can get your crops planted earlier. They also hold onto heat longer, so that you can get upwards of two extra weeks growing out of them at the end of the season. If you are raising a crop that can be harvested multiple times a year like lettuce, this can be enough time to fit another crop into the year. When growing in the ground, we can typically get two full harvests from lettuce. We can easily get three and sometimes even four harvests when growing the same crop in a raised bed.

So if you are looking to start a mini-farm, there is no better approach than to use raised beds. However, this assumes you have enough space outside in order to grow. There are some of us who don't have that space. While you might think that means you can't start your own mini-farm, this isn't true at all. You can still mini-farm even if all you have access to is a spare room in your house. You just need to use the right method. It is this method we turn to now.

Hydroponics

When it comes to farming, we almost always need to select crops that are native to our area or to an area with similar environmental factors such as sunlight, temperature, and humidity. That isn't automatically a limiting factor, as there are so many different kinds of vegetables, herbs, and fruits that we can raise in any environment. So, we still have a lot of variety, but we can't grow everything. At least, we can't grow everything when we rely on traditional farming methods or even raised bed gardens. Most of the time, when we talk about growing crops, what we are discussing is growing crops outdoors and thus relying on that environment. But hydroponics offers us a way of, not only fully controlling our growing environment, but also to grow our crops indoors. You can grow a healthy and bountiful crop in the comfort of your own home through hydroponics. But, with that said, you should keep in mind that hydroponics is also being used more often to raise full crops within greenhouses. This is excellent because it shows that we can scale up a hydroponic growing operation as our mini-farms prove to be profitable. Out of all the approaches that we have looked at, hydroponics is the most unique. We'll take a look at what it is, the different setups we can use, the pros and cons that come with it, and even how to quickly set up the simplest of hydroponic systems so that you'll see that it's easier than you might think.

Hydroponics offers a way of growing crops without using soil. If you have never heard of this approach before, then this might seem a little crazy. In traditional farming, you plant your seeds in soil. This soil has nutrients in it already, but we also fertilize our plants on a regular basis to make sure they don't run out throughout the growing season. In hydroponics, we don't use soil, and we don't use fertilizer. Instead, we use a nutrient solution. We basically just take a tub of water and then add liquid or dissolvable nutrients into it to create the perfect, nutritional blend for our plants. Our plants are then grown in an inert growing medium such as coco coir. This medium acts like soil in order to help our plants stay firm and supported, but it has no nutrients in it. Instead, the roots of the plants are allowed to grow out of the medium and dangle down into the nutrient solution. This may be achieved by having the roots dangle into the nutrient reservoir itself, or by having them dangle over a running stream of the solution. We may also use a top-down method to pump the nutrient solution over the inert growing medium so that it feeds through the medium to the plant's roots, and then back into the nutrient reservoir. This creates a closed system. Beyond the significant benefit of being able to grow plants indoors, this approach wastes less water than traditional growing methods, and there is less potentially harmful runoff.

One of the issues with hydroponics is that certain items are much more difficult to build and require much more money when compared to traditional crop farming methods. For example, when we grow a crop traditionally, all of the light is provided by the sun. Hydroponics requires us to purchase LCD grow lights in order to supply this for our plants artificially. That means we need to invest in lights and then continue to pay the power bill. It also means the lights have to continue to work. Some setups also require us to use an airstone to keep enough oxygen in the reservoir so that the roots of our plants don't just drown. Others need an air pump to push the nutrient solution through the various pipes so all of our plants can get their necessary nutrients. While this requires us to invest in these parts and use more electricity, they also create moving parts that can break down. If a part of a hydroponic system breaks down and isn't fixed quickly, you can easily lose a whole crop in a short amount of time. So when we look at disadvantages, we see that these systems require us to spend more money, devote more time to maintenance, and we also need to learn how to build them. With cons like this, you might think it is hard to recommend hydroponics, but it wouldn't be in this book if there weren't equally as many benefits.

One of the biggest benefits of hydroponics is obviously the fact that we can grow whole crops indoors through this approach. If you only have access to an inside location, then this is terrific. But if you have access to both an indoor location and an outdoor one, then you may still want to consider hydroponics as it allows you to increase how much you harvest each year. With traditional farming, we raise our crops during the growing season and then harvest them before the winter. During the winter, we are left waiting for warmer weather before we can get back out and sow the new crop. With hydroponics, since we entirely control the growing environment that a crop exists in, we can keep growing throughout the winter. Out of everything that we've looked at so far, hydroponics offers us the best way to increase our income because we no longer need to take a quarter of the year off. If you are growing for profit, this is amazing. But, if you are growing to feed your family, this is even better.

Another fantastic reason to use a hydroponic system is the fact that hydroponics produces larger vegetables and a bigger yield. A larger yield means more money, and so it should be clear why this is desirable. But often, when talking about traditional farming, we equate a larger yield to the use of chemical fertilizers. The use of these chemical fertilizers often leads to harmful runoff getting into nearby soil or waterways, as well as a reduction in the taste of the vegetables grown. Hydroponics is a closed system, and so there is no runoff. There is also no need to add any fertilizers to the mixture, as everything that the plants need is already in our nutrient solution. What's more, hydroponics helps us to produce tastier vegetables. There's been a lot of research done on this regarding lettuce, but the most impressive finding concerns herbs. Herbs can be a very profitable crop to grow, and when grown in a hydroponic system, they are up to 30% more aromatic. This means they smell and taste much stronger than herbs grown in the soil.

Hydroponic systems also rarely have to deal with pests or diseases. Most agricultural diseases are soil-based, and since hydroponics completely removes soil from the equation, those diseases are gone as well. Pests can still be a problem, but since most hydroponic systems are built indoors, it isn't simple for pests to get at them. To make sure this remains true, farmers need to be mindful to wash their hands and not bring outside pathogens into their growing area by accident. Farmers also need to remove any dead plant matter that's fallen off from their crops as this creates a place for disease and pests to grow. So long as the safety and cleanliness of the growing space are maintained, pests and disease shouldn't be a problem. You do need to be careful to ensure that your nutrient reservoir is closed properly, as water-based diseases might be an issue. However, these diseases are easy to avoid with just a little bit of consideration and an awareness of hygiene.

So, we've referred to systems in the plural rather than the singular. This is because there are several types of hydroponic systems that you can create and use. Which system is best suited for your needs is going to be determined by what you are trying to do. You can get away with a very simple system if you are growing a crop like lettuce, but you'll want a more complex system for raising tomatoes. You also need to be aware that hydroponics isn't a great match for every vegetable. Potatoes and root vegetables will still need to be grown directly in soil instead. We close out this chapter by taking a quick look at the six most common hydroponic systems, and then see how easy it is to build the seventh system, which is mentioned less but is a great fit.

The six main hydroponic systems are the nutrient film technique system, the deep water culture, the wick system, the ebb and flow system, the drip system, and the aeroponics system. There is also what we call aquaponics, which is a hydroponic system that uses live fish to create the necessary nutrients for our plants. Aquaponics is beyond the scope of our conversation, but if you are looking to maximize your use of space, then it can allow you to raise fish for sale alongside your crops.

A nutrient film technique uses an airstone and water pump to push the nutrient solution through a tube to wash over the roots of the plants. The plants are grown in baskets with slits that allow the roots to come out, and the baskets are placed in an angled tray. The tray is angled so that water flows back into the nutrient reservoir. This method washes the roots with water, but doesn't keep them in it or feed water through the growth medium. This makes the nutrient film technique the polar opposite of the deep water culture. The deep water culture has the plants grow in the same type of basket, but now the basket hangs directly into the nutrient reservoir. Normally we would be worried about the plants drowning, but an airstone provides the roots with plenty of oxygen.

Both the wick system and the ebb and flow system introduce the nutrient solution directly into the growing medium. The wick system is among the easiest systems to make, as wicks made out of nylon are run between the nutrient reservoir and the growing medium. The wick method is best used for crops like herbs or lettuce. The ebb and flow system works on a timer so that every so often, a water pump turns out and pumps water over the growing medium. The water then slowly drains through the medium and back into the nutrient reservoir. The drip system works like the ebb and flow system, but instead of flooding the growing medium, it slowly drips nutrient solution on top throughout the day and night.

The most complicated set up out of these is the aeroponics one. The plants hang down into the reservoir, but the water level is low enough so that the roots don't dangle directly into it. Instead, a water pump is used to push the water through nozzles that spray mist onto the roots. Out of all the systems we looked at, this is the most complicated and prone to breaking, and I do not really advise this for beginners.

If you are looking to start growing crops in a hydroponic system, then consider starting with lettuce in a Kratky method system. It's similar to the deep water culture in that the roots of the plant dangle down. But we typically refill a deep water culture, and we need to use an airstone to introduce oxygen to the roots. The Kratky method allows the roots to dangle down, but it is designed to be built and then left alone. The roots of the plants suck up the nutrient solution. Rather than filling the reservoir up, the roots are allowed to drain it so that there is more open air for them to get oxygen. By the time all of the nutrient solution has been absorbed, the plants are ready for harvest. All you need to set this system up is to take a container and fill it with nutrient solution, then cut holes in the top so that you can stick a mesh plant container into it. This method can result in lettuce that is 30% larger than traditionally grown lettuce, and yet it is as easy as can be.

What Approach is Right For Me?

If you are unsure of which method to use to grow your crops, there are a few simple questions you can ask yourself.

Are you growing indoors or outdoors? If you are growing indoors, then you will have to select the hydroponic approach.

Are you growing in the ground directly or in a raised bed? I suggest using raised beds whenever possible, but this isn't always the case.

Once you know the answer to the previous question, you can more easily figure out if you should use intercropping, mixed cropping, or monocropping. If you are growing in the ground directly, monocropping is out.

To choose between mixed cropping and intercropping, you should first set out your goals with your mini-farm. If you are looking to keep yourself fed, then you will probably want to go with a mixed cropping approach. If you are looking to earn money, intercropping is best.

Remember to rotate your crops every year if you are growing directly in the ground. If you are growing in raised beds, monocropping a profitable plant is possible.

Once you have the answers to these questions, you can start to plan your mini-farm and get to researching which plants make for the best combinations. If you still aren't sure what you should be growing, stick around for the next chapter where we look at profitable crops.

Chapter Summary

- There are many different approaches that we can take to raising crops. Each has its pros and cons and should be considered carefully.

- Monocropping is the act of growing the same crop in the same field, again and again.

- Monocropping is the most harmful of the practices that we can use to grow. It drains the health of the soil and requires chemical treatments to promote growth.

- People that argue for monocropping point out that it is more profitable because you don't need to purchase any extra equipment.

- Monocropping is so harmful to the environment that multiple world governments have outright banned it in the agricultural sector.

- Monocropping increases the risk of pests and disease.

- In contrast to monocropping, crop rotation is the act of planting a different crop in your field each year. Doing so allows the soil to stay healthier, and certain crops are even good at returning nutrients to the soil.

- Crops should be rotated on a three to four-year basis so that there are two or three different types planted before you get back to the original crop.

- Make sure that the crops you plant are all in different families. Planting two vegetables from the same family will defeat the purpose of crop rotation.

- Crop rotation takes more time than monocropping, but it is much healthier and results in bigger yields and better soil.

- Mixed cropping is often confused with intercropping, as both focus on growing more than one crop in the same field.

- Mixed cropping grows both crops in the same row, while intercropping grows two crops next to each other in close proximity.

- Mixed cropping is primarily used to increase the chances that one of the crops survives to harvest. It results in a smaller yield, but it can prevent a farmer from having nothing at all by winter.

- Plants are chosen for mixed cropping because of their differences. If one crop has shallow roots, then you plant it with a crop that has deep roots. If one crop likes dry conditions, you plant it with a crop that likes wet conditions. This helps to ensure that one of the crops makes it to harvest.

- Mixed cropping results in less issues with pests, disease, and weeds. This method also helps to promote healthier soil, too.

- Mixed cropping isn't as good if you are concerned with profits rather than consumption.

- Intercropping grows crops nearby to each other and offers almost all the same benefits as mixed cropping. The major difference is that intercropping results in larger yields.

- Intercropping is a longer process than mixed cropping, as you need to be careful when sowing the crops.

- Make sure that you carefully research the needs of the plants you are putting together. You want mixed crops to have contrasting needs and intercrops to have similar needs.

- Raised garden beds offer the best way of maximizing your space, and you should definitely be using them if you are mini-farming.

- Raised beds are garden beds that have been lifted off the ground by several inches to several feet. You then fill the raised bed with healthy soil, and it lets you have a much greater level of control over the growing environment.

- Since we control the soil in our raised beds, we could monocrop with them if we wanted to and not have to worry about damaging the soil.

- Raised beds are shut down for the winter with a layer of compost that adds healthy nutrients back into the soil for the following spring.

- Making a raised bed garden is time-consuming and the biggest downside to using them. You need to build them securely and make sure to use materials that won't poison your crop.

- Raised beds have fewer problems with soil degradation, weeds, pests, and disease.

- Raised beds can also be designed to be multi-level, and this lets us grow twice as many vegetables in the same space.

- Hydroponics uses a nutrient solution rather than soil to grow plants. It is also the only way for us to grow crops indoors.

- When herbs are grown in a hydroponic system, they are 30% more aromatic. Leafy greens like lettuce also show a roughly 30% increase in yield when produced hydroponically.

- Hydroponic systems cost more money than any of the other approaches, but they offer us the most control possible. Being indoors, they also don't have issues with pests very often.

- We can easily scale up a hydroponic operation, and there are enough of a variety of systems that we can choose an easy one to begin with.

- Potatoes and other root vegetables can't be grown in a hydroponic system.

- To decide which of these approaches is right for you, consider the space that is available to you to grow and what your goals are.

In the next chapter, you will learn all about which crops are the most profitable. Vegetables like potatoes, tomatoes, cabbages, and onions will be discussed to see how they are grown, harvested, and processed either for storage or sale. What's most profitable for you to grow will depend on the supply and demand in your local area, so consider visiting a local farmers' market to find out what is best for you.

CHAPTER THREE

PROFITABLE VEGETABLES TO GROW

In this chapter, we're going to learn what it takes to grow some of the more profitable vegetables there are. We'll also be taking a moment to discuss why you should consider adding herbs to your mini-farm, as these little plants are surprisingly profitable. Keep in mind that knowing what plants are most profitable for your mini-farm depends on the market value in your local area. The simple economics of supply and demand are at play here. If your local area has an abundant supply of potatoes, they aren't going to be nearly as profitable for you as they are for someone else. To get a sense of what is being offered locally, head to your nearest farmers' market and see what is being sold. You may want to take a pen and paper with you so you can keep track of everything you see. Write down what vegetables are being offered, how large of a supply there is, and what they are selling for. You can use this information to figure out if you can earn a profit selling alongside these other farmers, or if you should specialize in a crop that is under-represented.

We'll take a look at potatoes, tomatoes, cabbages, and onions in this chapter. We'll see what conditions they need to grow, how to ensure a healthy crop, and how to make money selling them down the road. Herbs will also earn a special mention in this chapter, along with the way that we can expand what we offer for sale without having to expand the crops we plant. This little trick might seem like common sense when you hear it, but you'd be amazed at how many farmers don't consider this simple way of expanding their products and income.

Growing Potatoes

Potatoes are an excellent crop because they lend themselves to many different recipes. They can be baked, mashed, fried. You can eat them as a side dish to most meals, or you can add some spices and fry them for homemade chips. They are such a staple of most people's diets that there is almost always a demand for them. They're also incredibly filling, which makes them a good choice for those whose mini-farms are primarily set up in order to feed their families. Potatoes are quite easy to grow and can net a decent profit, so long as you are careful when it comes to harvesting. Keep in mind that while you can grow potatoes in a raised bed garden, they aren't fit for being grown in a hydroponic system, and so those indoor farmers reading this might want to skip over this section. A vegetable like tomato, lettuce, or cabbage is more suited for the indoor approach.

Potatoes like to be grown in cooler weather rather than hot weather. However, they still go into the ground around the middle of April, after the last frost of the year has happened. This is important, even if you are using a raised bed garden. A frost can kill off your plants before they get a chance to begin, so never plant before the last frost. If you are absolutely desperate to get them started earlier, then you can risk planting them and use a tarp to protect them from frost, but this doesn't guarantee they'll survive. Rather than worry about starting mid-April, use a thermometer to check the temperature of the soil, and use this as your basis for when to plant. You want a temperature of at least 50F before you go ahead and sow your seeds.

Potatoes should be grown in soil that's loose, as they need to push their way through it. As potatoes grow, they'll push the soil away from them. If your soil is too tight, they won't be able to achieve this, and you'll end up harvesting tiny potatoes. Make sure that the location gets a minimum of six hours of sunlight out of the day. When planting in rows, they should be three feet apart from each other. Those planting in raised beds will be able to get away with a much smaller amount of space between each.

Potato seeds like to have about a foot between each other, and they like to be covered with a couple of inches of soil after they have been planted. Remember to water the seeds immediately after planting. If the conditions are right, then you should start to see seedlings poking through the soil after about two weeks. As potatoes grow, it is important that you do what is called hilling. Around the sprouting seedlings, parts of the potato will be visible at first. Push the soil up and around the seedling to cover this part. This creates little hills of soil around each potato plant. It's a necessary process since sunlight could cause damage to the crop and produce inedible potatoes.

If you are growing small potatoes on purpose, then you can expect to harvest three weeks after the seedlings begin to flower. For fully-grown potatoes, you want to wait until the foliage on top of the soil has fallen over and died. You might see this and think there is a problem with your crop, but it means they are fully-grown and almost ready to be pulled from the soil. Give them two weeks before harvesting, as this allows the skin of the potato to harden up and take on the texture that we associate with healthy potatoes. Keep in mind that waiting too long will result in a loss of the crop, so never wait more than two weeks after the foliage dies. While you usually water potatoes once every five days, you should slow down your watering as it gets closer to harvest. Reducing the water allows the potatoes to grow more firm and solid, the way that we expect potatoes to be.

When it is time to harvest, pick a dry day, and be extremely careful digging into the soil. Potatoes can be anywhere in the soil, as they have small cords that connect them to the foliage on top. If you were growing carrots, you could see that the carrot is directly connected to the foliage. But since potatoes aren't, they are able to spread out in the soil, and so often they aren't directly beneath the foliage at all. When digging into the soil, it becomes very easy to suddenly chip into a potato and damage it to the point that it can no longer be sold. Make sure that you bring your potatoes in from the sun, as this can cause them to taste to turn green and take on a bitter taste. You will want to store potatoes for at least two weeks in a location with a temperature between 45F and 60F. After they have been given enough time to cure like this, you can take them to the market to sell. This curing will let them stay fresh for a longer time, so you don't have to worry about your crop rotting before it sells. Just make sure that you never store them in a refrigerator, as this will shorten their lifespan. Also, don't wash your potatoes but instead use a brush in order to remove the dirt. Washed potatoes also have a shorter lifespan, so only wash potatoes when you are using them in a meal.

If you follow these steps, then potatoes can last a decent amount of time, and this makes them one of the crops that are great for starting out. When you are beginning, and just judging the interest, you might find that potatoes aren't in high demand and they don't sell as quickly as you expected. Since they can last a long time, you are more likely to sell them all off in the long run when compared to another vegetable, which only has a short shelf life.

Growing Tomatoes

Tomatoes can be a riskier crop. They require a lot more upkeep when compared to potatoes. They also don't keep as long in their natural form, and so it is a good idea to confirm there is a demand for them prior to planting. With that said, tomatoes also offer a way of creating more products down the line. We'll look at using tomatoes in this way later in the chapter.

It takes between 60 and 80 days for tomatoes to grow,

which means they have a long growing season, and this can reduce the overall profit you make. For example, you could grow two batches of lettuce in this same timeframe. However, tomatoes tend to sell at a higher price than lettuce, and so it can be worth it. But if you run into a problem with your tomatoes, you are going to have lost a lot of time.

Tomatoes should be grown in a location that gets full sun. They should have a minimum of six hours each day and a maximum of twenty-four. That's right, tomatoes like the sun so much that they could soak it up all day long if they wanted. This is worth noting because we can grow tomatoes in our hydroponic systems; however, they would eat up a lot of electricity. If grown in soil, make sure the soil is loose so that it drains quickly and that it has a pH level of 6.5.

Tomatoes are grown in a grid-like pattern rather than in rows. Plant tomatoes two feet apart from each other in all directions. Tomatoes grow on vines, and so we have to provide them with a trellis to prevent them from just dragging across the dirt. Set these up when you are planting, so that you can start to use them as soon as the tomatoes need them. You can expect to see seedlings growing up after a couple of weeks, but most people start theirs indoors and transplant them into the field after about two months. You will want to prepare the soil to have lots of phosphorus to grow big and beautiful tomatoes. However, keep the nitrogen level of the soil lower as they don't enjoy nitrogen very much and will quickly burn themselves out on it if you're not careful.

Tomatoes need to be watered more often than most vegetables. Make sure to water them deeply so that it promotes a healthy root system. Tomatoes should be watered right up until the time they are ready to harvest. Unlike many vegetables, tomatoes can be kept on the vine for an indefinite period of time. You'll want to harvest before the first winter frost, but otherwise, it is best to leave them on the vine for the longest period possible. You'll know they are ready to harvest because they'll have the bright red color of a healthy tomato while also being firm when squeezed. Some tomatoes may fall off before this happens. If this is the case, pick up the fallen tomatoes and put them into a bag in a dark room with a cool temperature. This can help them to continue to ripen and reduce the amount of waste the crop produces. To harvest, simply pick the tomatoes off the vine.

Make sure that you don't put new tomatoes into the refrigerator. Doing that increases their lifespan, but it reduces their flavor. Instead, try to sell your tomatoes as quickly as possible. If they aren't selling quickly enough, then stick around for another way to make a profit from this crop.

Growing Cabbage

Cabbage, like lettuce and other leafy greens, is among the healthiest options we can pick to consume. It also lends itself to multiple harvests in the growing season, and that means a better chance at making a solid profit. Cabbage also lends itself to hydroponics and, when grown by that method, shows a major increase in size when compared to traditionally-grown cabbage. It used to be a tradition in Ireland that you would plant cabbage on St. Patrick's Day. While we can start cabbage then, it is better to start cabbage seeds indoors and wait three or four weeks before transplanting them outside. That allows you to begin before the last frost so that your transplanted cabbages can get a head start growing. Anytime you can get a head start, that means a larger possible income when harvest comes.

Cabbages are hungry plants, which means that they soak

up all of the nutrients in the soil very quickly. You are going to need to fertilize cabbage regularly; otherwise, they will suffer from stunted growth. When transplanting, you typically want to leave a foot or two between each plant. How much room you leave is dependent on the size of the cabbage you are looking to grow. If you are looking to produce smaller heads, then you can plant them closer together and be ready to harvest them earlier. But larger heads require more space so they can stretch out and develop as they should. Keep in mind that even small heads of cabbage should be kept about a foot apart. This is due to the way, when they're too close, they compete with each other, eat up all the nutrients, and the result is smaller yields. If you are growing regular-sized cabbages, then you can expect two harvests in the year. If you are growing smaller ones, you should be able to squeeze in a third harvest.

To grow cabbages, you are going to want to feed them an NPK balanced fertilizer for the most part, but three weeks after they are in the ground, you also need to add some nitrogen into the soil. Nitrogen promotes foliage growth, and, since leafy greens are composed of foliage, nitrogen is the Holy Grail of nutrients necessary for large plants. You can expect to apply the NPK balanced fertilizer on a weekly basis. If you are growing full-sized cabbage, then expect it to take ten weeks before harvesting. Each plant should harvest about two pounds worth of product.

Harvesting cabbage should be done when it is dry out. Take a clean knife and carefully cut away any yellow leaves. Green leaves should be kept on, even if they are loose and don't seem very appetizing. These leaves help to keep the cabbage fresh while you store it. If you don't feel comfortable using a knife, you can always remove the full plant from the soil, and then hang it in a moist, dark location that is close to freezing. If you remove the heads properly, however, you only need to keep them in the shade. You can get a second crop, without having to sow any more seeds, by removing the head of the cabbage but leaving the leaves in place. This encourages the plant to grow a new head. These new heads will be much smaller than the original and are, what we call, microgreens. They aren't as profitable, but they can be just as delicious. When you are entirely done with the crop, it is important to remove all of the plant matter and the roots from the soil. If you neglect this, they will only promote the growth and spread of disease.

Cabbages should be wrapped in plastic and stored in the refrigerator. They can be expected to last two weeks this way. If you have a dark, dry location in your house like a cellar, then you can keep cabbages fresh for close to three months. Cabbages can be a very profitable crop, but it is crucial that you sell them quickly. You don't want to end up with a lot of rotting heads and no choice other than to throw them away.

Growing Onions

An easy crop to grow, onions might make you cry when they're peeled, but they'll make your bank account smile as you sell off this profitable crop. We start to grow onions when the soil is 50F, or the seeds we sow won't germinate and take hold. Since onions primarily grow in the soil, we want to make sure that we use a loose and quick-draining soil the same way that we do for potatoes. Onions are another vegetable that benefits from adding some extra nitrogen into the soil near the start of their lifecycle. With onions, you can add nitrogen at the same time as you sow their seeds. It is important to practice crop rotation with onions, as they usually deplete the nutrients from the soil. They aren't quite as bad for it as cabbages are, but they are certainly very hungry plants.

When planting onions from seed, keep in mind that their seeds do not live very long. You want to make sure you only purchase the freshest of seeds and get them into the ground quickly. Onions are generally grown half a foot apart in rows that are a foot to a foot and a half apart. Despite the fact that they primarily grow in the ground in the fashion of a root vegetable, producing onions has more in common with growing leafy greens. You'll want to provide them with a nitrogen fertilizer every couple of weeks, as this promotes large bulbs to grow. Once the onions begin to grow large, and the soil around them starts to fall away, you no longer need to fertilize them. Also, unlike potatoes, you don't want to create hills around your onions. When the soil falls away, let it stay that way.

Onions don't need to be watered frequently. If you have a layer of mulch on the soil, they'll need to be watered even less often. Typically, you can expect to water them about once a week. Just make sure that you don't forget. Onions are a little tricky, in that they could be going through a drought and still look perfectly healthy. But looks can be deceiving. Make sure that you set a watering schedule and don't let the dates slip your mind. You want your onions to *be* healthy and tasty, and not just look healthy while they are still in the ground.

As onions grow, you might notice that some of them start to send up flower stalks. When you see this happen, immediately harvest those onions. These aren't fully matured, but they have stopped growing. The onions harvested in this manner aren't going to be good to sell, as they only last a day or two. But you can consume them without any detrimental effects, so they don't have to go to waste. Mature onions have foliage that turns yellow and then falls over. When this happens, you can step on the stalks to speed up the process. Brush soil back away from the bulb, and then pull them from the soil once the tops turn brown instead of yellow. To store, you will need to remove the roots and cut the dry top off to about an inch from the top of the bulb. Onions can cure outside, just sitting on the soil, for a couple of days before being brought inside. Once inside, they'll still need to dry for about two weeks before being stored at around 40F. Onions can last for a long time, so you will be able to sell a crop for a few months after harvesting.

Growing Herbs

While there are too many herbs for us to look at each of them, it is worthwhile to note that these are among the most profitable crops you can grow. Basil, cilantro, chives, and ginseng are all among the most profitable crops that you can raise. If you are looking to earn money first and foremost, then you shouldn't neglect raising herbs alongside your vegetables.

Growing herbs doesn't take much work at all. You can and should grow them in a hydroponic system, as this will result in much more effective and appealing plants. Herbs are 30% more aromatic when grown hydroponically. While many hydroponic systems can be quite complicated to set up, herbs can be grown in a mini Kratky method system to great effect. Simply get yourself a small container with a tight lid, fill it with a nutrient solution, and place the herbs inside. Most hydroponic systems benefit from being grown indoors, but the Kratky system entirely cuts off the reservoir from the outside world, and you never need to open it. That means it works quite well when left outside in the sun. Make sure that the container you use isn't see-through, as direct sunlight hitting the nutrient solution will promote the growth of algae, and this will prevent the roots of the herbs from reaching the solution and getting the nutrients they require.

Herbs make for a wonderful addition to any mini-farm because they're quite small. You can easily grow a small batch of herbs without taking up any space at all. But if you are able to sell them and make a good income, it is simple to scale up a herb operation. You can also grow herbs all year round. Research which herbs grow the best in your local environment, and set aside at least one square foot of space to grow them. When you see how much money they can bring in, you'll be thankful you took the time to invest in them.

Jams and Salsas

We close out the chapter not on a vegetable, but one of the products that we can make out of them. Fruits and vegetables don't last nearly as long as we would like. If we aren't able to sell off our tomatoes quickly enough, then we either need to toss them out or eat them up. If we are growing primarily for an income, then this is a serious disappointment because both options mean we are making less money than we expected.

One way we can get around this issue is to make jams or salsas out of our vegetables. This doesn't work for every vegetable, but we can use some creativity to make delicious and natural mixtures like this to sell alongside our fresh vegetables. Plus, if you take the time to learn how to bottle a salsa, you can use that skill to process your vegetables and bottle them for longer storage. Jars of vegetables don't sell as well as fresh ones do, but jars of jam and salsa can be useful sources of income so long as you use a tasty recipe. What recipe is best to use will depend on your tastes, but don't be afraid to play around and try a bunch of different ones. This approach is primarily used to help us from losing more money on our crops, but it can also be a ton of fun.

Chapter Summary

- What crops are most profitable for you will depend on local supply and demand.

- Head to your nearest farmers' market and conduct market research to see what is being sold, how much it is being sold for, and whether or not there is enough demand for you to earn a profit.

- Potatoes are a profitable crop because they are a staple of so many meals and can be used for a wide variety of recipes.

- Potatoes like to go into the ground after the last frost in the early spring when the ground is around 50F.

- Since potatoes grow under the soil, make sure to use a loose soil, or they will have a stunted growth.

- Potatoes like to be planted about a foot apart from each other and then covered with soil. As potatoes grow, you need to push the soil up around the seedling. This is called hilling, and it prevents the sunlight from messing up the potato and ruining it.

- Potatoes can be harvested three weeks after the foliage begins to flower. This produces small potatoes. For full-sized potatoes, you should wait until the foliage has fallen over and died. Two weeks after the foliage dies, that's when you harvest.

- You need to be careful when harvesting potatoes as it can be easy to damage them in the process and reduce your profits.

- Harvest potatoes on a dry day and keep them for a few weeks in a cool location to allow the skin to harden properly.

- Tomatoes need at least 6 hours of sunlight each day, a pH level of 6.5, and 60 to 80 days to mature.

- Tomatoes should be kept on the vine until they have fully turned red, and they are firm when squeezed. Tomatoes that have fallen off can be kept in a paper bag in a cool area to allow them to finish ripening.

- Tomatoes need to be planted with about two feet between them. Use a trellis or a cage when planting to offer support to the plant.

- Tomatoes don't keep very long, and storing them in the refrigerator reduces their flavor.

- Cabbage is a leafy green that you could harvest several times during the growing season.

- Cabbage sucks up a lot of nutrients out of the soil, and it will starve other plants it is grown with.

- Add extra nitrogen to your soil for the biggest cabbages.

- Harvest cabbages when it is dry out, cutting away any yellow leaves, but keeping all the green ones.

- Cabbage, like tomatoes, need to be sold quickly; otherwise, they will go bad.

- Onions are a fantastic crop to grow for profit. They keep for a long time and are easy to grow.

- Make sure to water your onions on a regular basis, as onions going through a drought will look healthy even when they aren't.

- Don't hill the soil around an onion; the bulbs need to be exposed to the sun to grow properly.

- Among the most profitable of all crops are herbs. These small and easy to grow plants, net significant profits, and they benefit the most from being produced in a hydroponic system.

- Jams and salsas offer farmers a way of taking their vegetables that were going to go bad and creating a new product out of them. Selling jams and salsas can be very profitable, and they also ensure that we don't let any of our harvest go to waste.

In the next chapter, you will learn all about raising specialty livestock for profit. Cattle, chickens, goats, and even bees are just some of the many animals which can earn a high return on investment and be raised in mini-farming conditions. We'll discuss the needs of each and the expectations that are reasonable to have going into livestock mini-farming.

CHAPTER FOUR

RAISING SPECIALTY LIVESTOCK FOR PROFIT

So far, we have mostly talked about growing crops. The vast majority of those who are going to start a mini-farm will be sticking to vegetables and fruits. But this isn't always the case. Raising farm animals can be a deeply rewarding experience that also happens to be profitable, and so we would be remiss if we didn't take a look at how we can raise our own. In this chapter, we'll be looking at how to raise cattle, chickens, goats, and bees. While it should be clear that one of the creatures on that list is not like the others, that difference isn't reflected in how much income they can put in your pocket.

Raising Cattle

Cattle can bring a lot of money into your farm, but they are also among the most expensive types of livestock that you could raise. You can easily expect to spend up to $20,000 getting started in a venture of this kind. This is exactly the easiest thing to do when talking about mini-farming. Thankfully, it is possible to start a little smaller, but you need to be extremely careful and mindful of what you are doing if you expect it to earn a profit.

To begin, you are going to want to get your hands on some healthy cattle. Whether you want cattle for beef or cows for milk, you absolutely must first focus on ensuring that they are healthy. Don't purchase cattle without first inspecting them, as this is how you figure out what condition they are in. You want your cattle to be alert and interested when they meet you for the first time. Yet you don't want them to be wild, as this is a bad sign. Make sure that you look into the eyes of the cattle you are considering purchasing. See if there is anything leaking out, as this is a sign of sick cows. Take a moment and place your hand on the cattle and listen to them breathing. There shouldn't be any wheezing or coughing, and the breaths should come in regular intervals. Pay attention to how they move and make sure that they are full-bodied, as a thin cow isn't a healthy animal.

Next, you are going to need space to raise them. You want to make sure that they have an area that they can graze. But you don't want them to get out and run away, so you should be able to put up a four-foot fence around the grazing area. You also need it to be big enough to fit a shelter so they can get out of the rain or snow. Finally, there should also be enough space for them to hang out and relax, away from the section where they graze. That means a lot of space necessary to get started. But, once you have the space, you can continue to use it year after year. While you are considering their physical needs, also remember that you are going to need to be able to transport them. You'll want to obtain a trailer designed to haul cattle, and you may consider either renting or purchasing one, depending on how much you are expecting to use it.

New cattle are going to be quite stressed out, as cows aren't used to being taken for a drive. Try to offer them patience and understanding, talk to them in a quiet and calm voice as they leave the trailer. Don't rush them or yell at them, as this is likely to make them want to stay in the trailer and hide. New cattle should be kept in a smaller location to begin with, just long enough for you to check them over for signs of injury. By reducing the size of the location they are first kept in, you make it harder for them to escape. They are less likely to escape once they have calmed down. If you have a barn, let them stay in it for a day or two. Keep in mind that this quarantining doesn't account for disease. You should research to see if there are any cattle-borne diseases in the area, and, if so, you might need to quarantine your new cow for upwards of 100 days. Of course, if this is your first cow, then you don't need to worry about this yet.

While cows need space to graze, this doesn't replace their diet. You still need to provide them with dry feed, and you might be shocked at how much of the stuff they can eat. Cows eat a lot of food, so be prepared. The grass they graze on isn't actually to feed them, but to help them to digest their food and keep their bowels in working order. Along with the feed, make sure there is always fresh water for them to drink, and you might want to consider adding vitamins or minerals to the water to keep them as healthy as possible. There are also many feeds that have been fortified in this manner. If your feed is fortified, then don't worry about adding anything to the water.

You should brush your cattle on a daily basis. You don't need to go over every part of them, but rather just brush away anywhere that is particularly dirty. This isn't done quite so much with the intention of keeping them clean; it's more to do with keeping close to them so that you can spot any signs of illness early. If any of your cattle are sick, then you are going to want to check their temperature, heartbeat, and breathing rate to help you diagnose what is wrong. You should also pay attention to how much they eat, as cattle tend to eat less if they feel unwell.

That's all there is to it. You need to provide food and water for your cattle, make sure that they aren't getting sick, and take them to the vet when they are. But beyond this, it is just a matter of time until they are ready to be sold. Or, if you have the stomach for it, you can also butcher your cattle yourself to sell the meat. But this process is much more involved, and many of us simply don't have the constitution for it. Selling fully grown cattle can earn you a tidy profit, so there is no need to worry about getting your hands dirty anyway.

Raising Chickens

Raising chickens can be a bit of a hassle. If you've ever had to go chicken catching, then you know exactly what I am talking about! But despite this, they are one of the more easy and profitable animals to raise. You can keep them in your backyard without a fence even, and you can get eggs and meat from them. Chicken feces makes a great fertilizer, and the shells of eggs can be added to the compost pile to make a nutritional mixture for your crops. Chickens will also eat worms and other insects, so they can serve as a way of reducing the number of pests you have to deal with.

Before you purchase any chickens, you will want to build a coop. A proper coop has space for a feeder and container for water. It must also have a roosting area and at least one nest box. If you only have a couple of hens, you will only need one nest box. But you should have three nest boxes for every ten hens you have. Keep in mind that you are also going to need to be able to get into the coop, as this is how you check for and retrieve any eggs that have been laid. Make sure that you build your coop out of a solid material, as nothing is more horrifying than having to clean up a collapsed chicken coop. Keep in mind that the size of the coop is going to be determined not only by how many chickens you have but also what breed they are. Typically, you shouldn't have less than three square feet inside of a coop. Chickens will benefit from more space, so consider going larger but never consider going smaller.

But before you even worry about building a coop, make sure that it is legal for you to raise chickens. Some towns don't allow any chickens to be raised on some kinds of private property. Others set a limit on how many chickens you can have. Make sure that you know what you are allowed to raise before you start investing your time and money into building a coop that you can't use.

Chickens need to be fed daily, but they don't eat as much as cows. Chicken feed is also much cheaper than cattle feed, so you don't need to spend much money to keep them fed. Purchase a 50-pound bag of feed and use that to calculate how much you are going to need to invest on a monthly basis. If you only have a couple of chickens, then this could even last you the whole month. But the more chickens you have, the quicker they're going to get through the bag.

While you can sell the chickens themselves for their meat, it tends to be more profitable to focus on selling eggs. Hens typically lay eggs in the spring and the summer. If you live in an area that has plenty of sunlight during the fall, you can expect them to continue laying eggs throughout this season too. Hens lay eggs so often that you should make it a habit to check for new eggs once in the morning and once before bed. Keep in mind that chickens are almost like dogs and so you'll want to have somebody coming over and checking on them during the day if you have to be away from home for a bit. Make sure you ask them to check for eggs and that you warn them about how much chicken manure there is inside the coop.

You should never purchase only a single chicken. These animals are very social, and they like to be kept with others of their kind. You shouldn't purchase less than three, and you are better off getting five or six if you are legally allowed to. Since the average chicken lays an egg every day and a half, you can expect to get roughly an egg a day. As chickens get older, they will start to produce less. Typically this happens after the second year of life, at which point you may want to consider selling the bird to the butcher and purchasing a younger bird to keep the egg production higher.

Chickens that are crammed together are more likely to cause and spread disease. Chickens also need to get exercise, so they should have enough space around the coop to get out and spread their wings and run around a bit. Be aware of the possibility of predators, and take steps to ensure they can't get in. While wild animals like foxes are particularly associated with chickens, don't forget that your dog or cat will also rip them to shreds if given half a chance.

Where cattle can easily run you a few thousand dollars for an initial investment, chickens are a much cheaper animal. Building your coop will be the most expensive part, and this can easily cost you $500 or more. But chickens typically don't sell for more than $50 and can often be purchased for as little as $5. While you should be cautious about chickens being sold for such a low price, it isn't necessarily a red flag. Chickens breed like crazy, and so they often end up being offered at low prices simply so the owner can move the stock.

To make money from your chickens, you should primarily be selling the fresh eggs. Three chickens produce about seven eggs a week. So there's no reason you can't sell a carton of eggs each week. If you only sell them for $3, you can make back a $600 investment in 100 weeks or less than two years. At the end of the second year, say you send the birds to the chopping block and purchase three new ones. Even if you paid $50 a bird, it would only take you 50 weeks to make back the investment, and you could make another $150. If you purchased birds at $5 each, then you would make that back in 5 weeks and have 99 weeks left to earn profit. Of course, they are going to take the winter month off from laying eggs, but you can see how chickens can quickly start to pay for themselves off their eggs alone.

Raising Goats

Goats might not be the first animal you think of when it comes to livestock for your mini-farm, but they can be a surprisingly profitable investment. Goats can be raised for meat, but to get the most money out of them, you should keep them for their milk. They produce ten months out of the year, and you get close to 100 liters of milk from them each month. Not only that, but studies have shown that goats bond with humans just as strongly as dogs do. This means you can have a profitable milk-machine and a new best friend. As with cattle, you must prepare your property for a goat prior to purchase.

The preparation of your property doesn't need to be quite as extreme as with cattle, but you are still going to need to make sure you have a fence to keep them from wandering around the neighborhood. Goats are a tasty prey for large cats and dogs, so the fence doubles as a way to keep them safe. You'll want to be able to divide up the inside of the fenced area so that there is a location from them to shelter, another for them to eat, an area for milking, and an area for babies if you decide that you want to breed goats. Out of these areas, the shelter is the most important. You could always milk the goats in the middle of the field, and you can always keep the food around the shelter. But without shelter to protect them from rain, snow, and other environmental factors, you will find yourself with goats that are sick a lot more often than you expected. The shelter should be packed with hay for bedding and kept comfortable and regularly cleaned. It helps to make it a more relaxing area for the goats, and they will be able to return here to avoid stress, which would otherwise lead to more health issues or even a reduction in the milk they produce.

You'll also want to purchase supplies before bringing goats home. You'll need containers to hold their food, a mineral feeder to ensure that they get enough nutrients, and a water trough or a large bucket. You may also want to get supplies for bathing them from time to time, a brush to keep their fur clean, and maybe even a collar and leash to walk them. It may seem silly to walk a goat, but they thrive on exercise as it helps to keep their body strong and healthy. Speaking of healthy, you are also going to want to take a few minutes and open up Google to see what plants are poisonous to goats. For example, azaleas, china berries, black cherry, Virginia creeper, and honeysuckle are but a few plants that will make a goat extremely sick. Get a list of poisonous plants and make sure that there are none within the area you've fenced off. Goats are like cattle in that they graze on plants constantly. They love chewing grass, leaves, and any greenery they can find. They'll happily munch away on poisonous plants without a second thought, so it is up to you to keep them safe from themselves.

The final thing you are going to want to obtain is a First Aid kit. This is useful for your cattle too, but goats are a little more energetic and prone to getting into accidents. Get bandages, clippers for trimming hooves, and a proper syringe for injections. It is also a smart idea to prepare yourself by looking at the warning signs that goats give off to let us know they're sick. Some signs are easy to spot like coughing, gunk leaking out of their eyes, or discoloration of the face. Other signs require you to watch their actions to see if they are grinding their teeth, avoiding meals, pushing their head into a wall, refusing to get up, or isolating themselves. Finally, pay attention to see if they are chewing their food properly, whether their feces has become runny, or if their udder is unusually hot. These are just some of the ways that goats signal to us that they need to see a vet. You can always have a vet visit your farm if need be, but you should also have a trailer to take them in yourself. To help out the vet, you should listen to the goat's heartbeat, take its temperature, and keep track of any physical signs such as discolored gums or problems with their feces. That will make it easier for the vet to quickly diagnose your goat so you can get treatment for the problem sooner.

While even the healthiest and highest quality goat can get sick, sickness is much less common when you purchase from good stock. To do that, there are a handful of useful questions that you should ask whoever you are planning to purchase from. The seller should have an answer for most of these questions. Not being able to answer one or two is alright, but if they can't answer any of them, then there is a problem, and you should avoid buying from them. Begin by seeing if their goats are registered, as this is a good sign of a trustworthy seller. Ask how often the goats are tested for disease and what vaccinations they've had. Ask if they've had any goats die from disease. If they have, this isn't necessarily a bad thing. Follow up and get the story of what happened from them. If they are willing to share the fact that a goat has died of disease, then this is actually a good sign of a trustworthy seller. A candid seller is always better than one that lies.

Next, you'll want to know more about how the goats were raised so that you can more easily match your mini-farm situation to their expectations. Ask them about what they've been feeding them, how much fiber they provide, and how much milk the goats have produced. Moving from the seller's farm to your farm will be a stressful experience for the goat, and so you can make it smoother by matching your feeding to what they're used to. Less stress means healthier goats.

If you are careful to keep your goats healthy, they can produce milk for up to two years, at which point you could sell them for their meat. If you decide to breed goats, then you will need to be more careful about milking. Pregnant goats should be left to dry up and stop producing; otherwise, it can risk their health. Breeding goats can be quite profitable, but if you are after goats for their milk, then you are better off avoiding breeding.

Raising Bees

You might not think of farming and bees in the same category, but there's no reason you can't add a few hives to a mini-farm. Bees don't take up very much space, yet they can be incredibly profitable. Before you even consider purchasing any, you should first check with the zoning laws in your local area to see if you're allowed to.

As with chicken, some areas don't allow bees to be raised on certain properties.

If you can raise bees, then the first step is to dedicate a space for them. Picking a location for your bees is a little like selecting a location for a crop. Make sure that they can get plenty of sunlight, though afternoon shade is also necessary. This means you should watch a chosen location and track how many hours of sunlight and shade it gets. Next, you will want to make sure you can provide nearby fresh water. Bees need water to stay healthy, and so you'll want to dedicate some water for them. If you don't, then you might find them coming into the backyard to hang out in your pool or birdbath. That might prove deeply unsettling for guests, children, or pets, and so you are best served by keeping fresh water as close to the hive as possible. When we keep bees, the hives are built into wooden structures. The wooden structure isn't necessary for the bees themselves, but they provide protection from wind and rain, which could otherwise damage their home. Bees like to be left alone as much as possible, and so you'll want to place them away from any areas that see lots of people or animals come through. If you have the room, you should place your hives 50 feet away from anything else. If you don't have the space to keep them at that distance, consider planting hedges or building a fence so that people don't disturb them. That has the secondary effect of keeping your guests, children, and animals from worrying about their presence nearby. Finally, you should keep the hives pointed to the south and raised off the ground. This will give them better use of the light while also protecting them from the elements and possible predators.

With your location prepared, wait until spring to get the hives put in. Bees are purchased in frames, which are then slotted into the prepared location. These frames are filled with honeycombs and are easily pulled out to drain the honey and check to see if the queen is laying eggs. The cheapest and easiest way to get started with bees is to purchase some packaged bees along with a caged queen, but this can take a long time to get a healthy and full-sized colony going. If you are starting out, then a more expensive but easier way is to purchase a nuc, which is a young hive made up of two to five frames of honeycomb. A nuc also has a young queen who has only just begun to lay eggs, which means that they'll be producing quite a few offspring at this point. It's important that you purchase a nuc from a reputable breeder. The other option that you have is to buy a full colony, though this is jumping into the deep end if you are a beginner. I would recommend starting with a nuc because it will be large enough to be productive quickly. Purchasing packaged bees will take longer to become productive, and beginners may feel like they aren't achieving anything with this slow start. By beginning with a nuc, you have the room to grow it into a full colony, and so you can be present for each step along the way. Then, after you have raised a full colony, you can start to think about expanding and purchasing another established colony to add to your mini-farm.

In order to install the bees into your farm, you should follow the instructions provided by the seller. If your seller does not have guidelines for how to install them, then they are not a reputable seller and should be avoided. This is almost the bare minimum that you should be able to expect out of a seller, so avoid anyone that won't offer this advice. They should be able to tell you how to install them, how to use the necessary tools such as a smoker, and they should have advice on how to work with the bees in a manner that won't stress out the bees or damage the hive. You will want to have a beekeeper's outfit, including gloves and facemask, as well as a smoker.

With your bees newly arrived at your farm, the first step to taking care of them is to raise them. Even though the frames you purchased are covered in honeycomb, your bees are nearly homeless when they are first installed. They need to build and make their home, as all you have provided them with is a frame. It might seem like it has walls and a roof to you, but your bees will feel differently. They are going to go through and seal up any cracks and start storing food. They're going to be extra careful to ensure the queen is safe, secure, and comfortable. To do this, they are going to need lots of food for energy. This food comes in the form of nectar. You might think this means planting a ton of flowers, but we can make our own nectar-like concoction that the bees will love. Take a jar and fill it up to the halfway mark with sugar. Fill the second half up with water and mix it around. You should have a viscous sludge. Use a feeder lid and store the jar upside down. The mixture should be solid enough not to leak through the lid. Bees will be able to drink from the feeder to get a sugar high that helps them build up the colony. In the first few weeks, you can expect them to drink nearly a jar of the stuff a day. But by the time a month has passed, they shouldn't need any more homemade nectar as they'll be heading out from the hive to find flowers nearby and feed on them. You can assist them by growing flowers in the area, but they'll find some even without your help. When you notice that they are doing this themselves, remove the feeder. The honey they'll produce using homemade nectar is nowhere near as tasty as what they produce when they make their own. In fact, you shouldn't sell the honey they produce in this early stage as it will only leave customers with a poor opinion on the quality of your goods.

As the hive takes shape, you'll want to be involved and watchful. Take a look at the hive at least once a week, as this will let you see how it builds and forms. Once you have a solid understanding, you can slow down and inspect once every other week or so. Keep an eye out for problems such as excessive bee feces, ant infestations, or problems with the queen, such as a reduction in eggs laid. You will also want to check for varroa mites and diseases like foulbrood. In time, the bees will grow too large for their frames and require you to add in some more. If they grow beyond eight frames, then a second box will need to be filled. Begin expanding by using a second box to support the original colony; the second one can be turned into its own colony eventually. Before you expand, however, you should think about protecting your colony by using an entrance reducer. That's only a simple block of wood that can help to keep pests and natural enemies out of the hive. When a hive is smaller, they're less protected. As the hive grows, the bees will have a higher defensive capability, and they can handle themselves just fine. You should keep in mind that an entrance reducer will also reduce the amount of honey you get. So, when the honey is at its most productive, you are better off avoiding a reducer.

If you have started your bees in spring, you might be able to get honey during the summer, but the chances are you're going to have to wait until the following spring. The bees typically need some time to prepare the hive before they start producing much honey. Expect to begin collecting late spring, the year after purchase. A typical frame will provide about eight pounds of honey. You can purchase specialized extractors to collect this, but beginners should learn how to do this on their own before buying expensive gear. One method is to use a scraper to cut off the honey from the frame into a container below. Don't worry about the beeswax; just collect it all at the same time. Next, use some cheesecloth or a strainer to filter out the honey from the solid bits of wax. This is left to sit until all the bubbles rise through it to the surface. From then, you can start to bottle and prepare your honey for sale. You'll find that the eight pounds of honey per frame has fallen to about three or four pounds after you removed the wax and let it settle. It might seem like you lost a lot of potential profit there, but there are ways to use this wax for larger profits.

Speaking of profits, we turn now to making money with our mini-farms.

Chapter Summary

- Raising livestock requires a much steeper initial investment, but it can bring a major profit in the long run.

- Cattle are the most expensive of the livestock we've looked at. They are expensive, and they require shelter, a large field, and lots of feed and water.

- You should always make sure that your cattle are healthy before you purchase them. Make sure to ask lots of questions before purchasing, Reputable sellers won't mind you asking, and they'll have honest answers.

- New cattle are going to be stressed out, so give them lots of quiet and give them some time alone. If they are your first, then they only need a couple of days. If you already have cattle, you will want to isolate them for up to 100 days.

- Cows need grass to graze, but they still get through a lot of feed and fresh water every day.

- Chickens are cheap, but they still eat a lot. They'll need a coop that is big enough to give them enough space to stay healthy.

- You should always check with your local township to see if it's legal to keep chicken on your property.

- Never purchase a single chicken; they are social animals, and you shouldn't have less than three at a time.

- Goats are as friendly as dogs, and they produce lots of milk. They aren't as expensive as cattle, but they still need a fence, room to graze, and a shelter to keep out of the elements.

- There are a lot of plants that are poisonous to goats, so make sure that none of them are present within the area you've fenced off for them.

- Make sure to only purchase your goats from a reputable breeder who will answer your questions about their health.

- Bees might not be what springs to mind when you think of farming, but they can be an extremely profitable investment.

- Bees will need a location with shelter from the wind and sun during the morning, but that's shady in the afternoon. They require water nearby, and a home where they won't be bothered by foot traffic.

- For beginners, purchasing a nuc is the best way to go with your first hive. It will be productive earlier but isn't so big as to be overwhelming.

- Create a mixture of sugar and water to help your newly-purchased bees settle into their space. They'll go through nearly a jar of the stuff a day, but after three weeks, you should be able to stop providing it. The honey produced during this period will taste much worse than normal and shouldn't be sold to customers.

In the next chapter, you will learn how to make a profit from your mini-farm. Farming may sometimes be a relaxing and enjoyable experience, but the end goal is still to turn a profit. To do this, you need to learn how you sell livestock, vegetables, eggs, honey, or whatever else it is that you've decided to cultivate on your mini-farm.

CHAPTER FIVE

HOW TO PROFIT FROM MINI-FARMING

The goal for most of us is to make money with our mini-farm. It may be enjoyable to plant or crops or tend our livestock, but if we don't turn a profit, then we can't afford to keep our farms going. Thankfully, there are many ways in which we can earn a profit from our livestock and our crops. In this chapter, we're going to see some of the many ways in which we can do exactly that. We'll look at how our cows, chickens, and goats can earn us money. We'll see how our crops are able to bring us a pretty penny, and even how our bees can earn their keep through honey and additional products. There are lots of ways to turn a profit with a mini-farm, and I hope that those we'll look at get your brain working; a little creativity can come up with thousands of interesting and unique ways of earning a dollar.

Profiting From Chickens

An investment in raising chickens takes a little start-up cash. After all, not only do we need to purchase the birds, but we also need to build them a coop. If you have built it soundly, then a single coop can last you several decades, with minimal repairs and upkeep. This is great because it allows you to make an initial investment that you can pay off quickly so that you can earn profits within a couple of years. As far as livestock goes, chickens are among the cheapest to get started. But they also don't earn a lot of money out of the gate, so you should consider them as a long term investment.

The most common way of earning money from your chickens is to sell the eggs they lay. The idea might seem silly when you go to the grocery store and see a few hundred cases of eggs from half a dozen brands. Yet there is still quite a bit of profit to be made through selling eggs. Why is this? The answer is surprisingly obvious when you hear it. Out of all the large scale farming operations, raising chickens is often the most disgusting and inhumane. These birds are housed too tightly to be healthy, and videos of their abuse have been leaked online from several different sources. These inhumane practices have affected people. Many have sworn off eggs and chicken products, but there are plenty who have looked for alternative sources that act in an ethical way. That's where you come in. If you followed the advice in the previous chapter, then you have created an ethical and humane chicken farm, and you can sell your organic eggs for $5.00 a dozen. With six chickens, you can sell five dozen eggs in a two-week period.

Furthermore, once the hens stop producing eggs, you can then sell them off as meat. Egg-laying chickens aren't going to sell for as much as birds grown for meat, but you can still expect to get a couple of dollars per pound. But if you decide to raise chickens for meat, then you can have a quick turn around, as it only takes about two months for a bird to be ready to hit the butchers. Depending on the quality of the bird, which will be determined by breed and health, you can make upwards of $6.00 a pound.

While you may think eggs and meat are the only things chickens are good for, you should keep in mind that they poop constantly. This may hardly strike you as a positive, but chicken manure is great for use in adding nutrients to the soil. You may want to keep your chicken manure for yourself to use in your fields, but chances are there's going to be a lot more than you can use yourself. With a little bit of forethought, you can package and sell bags of manure to local farmers or gardeners. This little bit of creativity means more money in your pocket as you literally turn poop into cash.

Profiting From Cattle

Out of all the livestock we've looked at, cattle are the most expensive to get going. They need more space, shelter, and food than chickens, goats, or bees. Because of this, it can be easy to lose money on your investment if you aren't careful to keep them healthy and in good shape. But if you are able to provide them with the necessary care, then you can earn money by selling their milk or products made with their milk. You can also raise them to sell as meat, or you can raise a strong bull for use as part of a breeding service. Plus, as with chickens, you can use their waste as fertilizer for your crops, or sell it to local farmers and gardeners.

As far as profits go, milk is the best one because it is repeatable. But it also takes the most work because you need to milk your cattle twice a day. The laws around whether or not you can sell that milk are going to be determined by where you live. Some areas put strict regulations around the sale of milk, and this can make it pointless for you to purchase milk cows in the first place. Other places have rules in place to allow for mini-farms to sell their milk, typically with a limit on how many milk-producing cows you can have before stricter regulations are required. With these small-scale licenses, it's very unlikely you'll be able to get your milk into grocery stores. Farmers' markets and locally owned organic food stores will be your target demographic in this case.

Now, there is one way in which you might be able to make a profit from milk cows even if you can't sell your milk. The trick is to process it and create daily products for sale. Homemade cheeses, yogurts, butter, and more are all possibilities. These products often still require some form of registration or license, but many can be sold locally at farmers' markets without any problem. This approach is a great way to earn money from your cattle as you can use family recipes to make unique products that don't have stiff competition the way milk does.

Selling your cattle as meat is another way of making decent money, but it costs a lot up-front. To make it worth your while, you need to be breeding your own cattle; otherwise, you would just be purchasing cattle to resell them afterward, and that is a recipe to lose money in the long run. You'll need to learn how to breed your cattle, how to tell they are ready to be sold, and how to ensure they are as healthy and tasty as can be. If you have the patience to go through the process of preparing cattle for sale as meat, then you can earn upwards of $6.00 a pound. That might be the same price as top-quality chicken, but there is quite a bit more weight to a cow than a chicken.

Finally, you can also offer bull services to local farmers and help them in breeding their cattle. A strong bull will make for strong calves, and this is something that every farmer desires. Yet not every farmer needs to keep a bull around, as many only need one when it comes time to breed. By purchasing or raising a strong bull, you can easily make over $100 a month by selling its services. If you are breeding your own cattle, then keeping a bull around can help you ensure that your cows are tended to when they are in heat. You might not notice they've entered, but the bull will, and he'll know exactly what to do. Out of all the ways of making money off cattle, this is the most profitable when you weigh costs to income.

Profiting From Goats

As with cows, goats produce lots of milk, and you can

sell this milk for a decent price. You will still run into issues regarding the legality of it as determined by your state, but regulations are often much more lenient than with cow milk. Unfortunately, goat's milk has a unique taste, and many people don't enjoy it. The reduced demand makes it a specialty item, which means that you can expect to make $2 to $5 more per gallon when compared with cow's milk. Some people have issues with digesting cow's milk or dairy, and so goat's milk is one of the alternatives that they turn to. If you have any specialty coffee shops in your area, then you should try to sell goat's milk to them, as they use a lot of milk and are most likely to offer these alternatives to their customers. Soy and almond milk are always popular, but local, fresh goat's milk is hard to beat and can make for a powerful marketing tool for these businesses.

If you are having a hard time selling the milk as it is, consider making products with it. You can always use it in different dishes, but one of the best ways to profit from goat's milk is to make bars of soap. Goat's milk soap is an anti-allergen, and so it is highly valued by those in the beauty industry. You don't need a license to sell soap. You can take it down to local markets if you want, but the best money comes from selling online. You can create a shop on Etsy for free, and offer goat's milk products that customers can purchase to have shipped to them. This opens up your mini-farm from primarily making a profit from the local area, and it taps into the power and interconnected nature of our modern society.

Eventually, your goats will stop producing milk. When that happens, you can always sell them for meat. Like goat's milk, goat meat is a specialty product. Therefore, it can be harder to sell, even more so than their milk. But if you have healthy goats of a decent weight, then you can net upwards of $200 selling them to the butcher or at the market. If you are breeding goats, this can be a great way to make a lot of money. If you aren't, then you can use this to purchase new goats for milking.

Finally, with a little creativity, you can earn upwards of $200 a day off your goats. If you have enough of them, they can be hired to help tend to lawns. Goats chew everything they can get their teeth on, so some people have turned this into a business and rent out their goats to serve as a natural lawnmower. It sounds a little odd, but it can absolutely make you a lot of money.

Profiting From Bees

Bees produce a lot of honey, which you can collect, bottle, and sell. You may want to sell this to local businesses, but you can also sell it at your local market or even alongside the road if you set up a booth. There are much fewer laws limiting the sale of honey when compared to the sale of milk. You can expect to make close to $50 off a gallon of honey. But remember that the honey we collect from the comb is very waxy and needs to be properly strained. You sell the honey that strains from that, but then you are left with a bunch of wax. However, this needn't be wasted.

Beeswax, beyond being none of your business, can be used to create many profitable items. One of the most common uses for beeswax is lip balm. Like goat's milk soap, you can sell beeswax lip balm online and make money from all over the world. Another product that is often made is hand salves. Beeswax has also been used in moisturizers, creams, eye shadow, blush, hair pomades, and other beauty products. If you keep your products natural, then you can make even more money through this approach; people are willing to pay extra for the knowledge that they are using natural products rather than those filled with possibly harmful chemicals.

Another way you can make money from your bees is to sell the honeycomb itself. Sold as a delicacy, the comb is a delicious snack that can be combined with meat and cheeses for a truly mouth-watering flavor. You can probably find local restaurants that are interested in purchasing your honeycomb. It has become more popular with the public in recent years, and it's also easier to sell honeycomb directly to customers rather than businesses.

Finally, to approach this with creativity, did you know that you can rent your bees out to farmers? Bees help pollinate flowers and plants, and this can help farmers to increase the size of their harvest. There are many crops that need to be manually pollinated, but even those which automatically pollinate themselves can see a major increase in the size of their yield when they use bees to help out. This unusual way of earning money often doesn't bring in a fortune, but the bees will keep producing honey at the same rate, and so you don't lose out. That means it's pure profit.

Profiting From Crops

What crop you grow is going to largely determine how much money you make off it. This number then changes based on where you live and sell and what kind of local supply and demand is in place. We don't need to dwell on this since it's been discussed previously. Instead, let's look at the three areas that crops are most commonly sold.

The first is the farmer's market, as we've mentioned repeatedly. Some markets require you to rent a table, but some allow you to set up for free. You'll need to provide transportation for your crops, and you'll want to bring enough money to give change to customers. Simply lay out your produce and let people buy them as they desire.

A more profitable arrangement is to sell your crops to local grocery stores. This can be a hassle with larger chains. They do often purchase from local suppliers, but they tend to have arrangements with larger providers, as the demand is too much for most mini-farmers. You are better off trying to sell to local organic stores or specialty grocery stores. They have a smaller demand, but they also have a smaller supply. Just make sure you grow your crops organically. If you don't, they're unlikely to want anything to do with you.

Finally, it may sound funny, but often the most profitable way to earn money off the crops you grow through your mini-farm is to set up a roadside stand. There might be zoning laws that prevent this, but these are pretty rare, as they are only usually active in urban areas with large populations. If you live in a smaller town, you're almost guaranteed to be able to do this. Which is great because it can make a lot of money. There is one such roadside booth that operates just down the road from me. The vegetables are left in baskets, and there is a money box chained to the booth. Each price is clearly listed, and the whole operation functions on the honor system, so it runs even while the farmer is out working the fields or taking his family to church. Even though it's such a simple system, it makes a lot of money. People claim that he grows the best tomatoes in the state, and people come from all over to purchase.

Remember, too, that you can always make salsas, dips, or sauces out of your vegetables if they aren't selling fast enough. You can also bottle and can certain vegetables so as to keep them for a much longer time, though these don't sell anywhere near as well as fresh vegetables. With a little creativity, you can probably come up with some amazing recipes that not only taste great but sell like hotcakes.

MINI-FARMING

Chapter Summary

- Chickens are best used to earn a profit from the eggs they produce, as organic eggs can be a hot ticket item.

- When your hens stop producing eggs, you can sell them for meat. You can also grow chickens to sell for meat if you have the space and inclination to.

- While cleaning out chicken poop can be annoying, you can also sell it as a nitrogen-rich manure to keep the local fields healthy. Or you can use it in your own fields.

- Cattle are great for producing lots of milk that can earn a tidy profit, but there are also lots of regulations around the sale of milk that you should be aware of.

- If you can't sell the milk from your cows directly, consider using it to create dairy products like cheese and butter and then selling these.

- Cattle can earn up to $6 a pound when sold for meat, and you can even rent out bulls to help other farmers breed their cows.

- Goats produce a lot of milk as well, but it is a specialty item that sells for more, though it has a lower demand.

- Goat's milk can be used in a number of beauty products such as soaps, and these can then be sold online rather than just locally.

- Goats can also be sold off for meat; this is best done when they've stopped producing milk.

- You can even rent your goats out as all-natural lawnmowers to help weed fields or lawns.

- Bees produce lots of honey. This can be sold for a good price.

- They also produce a lot of wax, and this can be sold for use in beauty products, or you can make the beauty products yourself and sell them online.

- Honeycomb can also be sold as a delicious and natural snack.

- Some people even rent their bees out to local farmers to help them pollinate their fields for a bigger harvest.

- You can sell your crops at the local farmers' market.

- Local grocery stores may also purchase your crops, though they are more likely to do so from larger farms that can meet their stiff demand.

- A small-scale but profitable way to sell your produce is to simply set up a roadside booth and sell them directly to consumers.

- You can also take advantage of canning and jarring to preserve your vegetables before they go bad, or you can mix them up into salsa and sauces for sale at the farmers' market.

In the next chapter, you will learn how to prevent pests from infesting your crops and ruining your livelihood. These annoyances can make your life a nightmare, but we are able to manage them by practicing proper maintenance techniques to ensure we are growing healthy crops and animals. More than anything else, proper maintenance leads to better profits.

CHAPTER SIX

PEST PREVENTION AND MAINTENANCE CONTROL

With our mini-farms, nothing could be more devastating than losing a crop to disease or pests. These annoying problems can be easy to treat when caught early, but if they are left untended, they can quickly wipe out your entire operation. In this chapter, we'll consider the ways that we can maintain the health of our mini-farms so that this never happens to us.

Maintaining Your Crops

For the most part, you aren't going to need to do much to maintain your animals. The biggest issues that you'll have with them are going to require the help of a vet, and so you will have much more direct and actionable information to work from when you run into issues. But your crops are entirely dependent upon you. This means that you need to know how to tend to them properly and keep them in good health.

To begin with, you want to make sure that you aren't overwatering them. Too much water promotes the development of root rot, and this is one of the deadliest diseases. The problem with root rot is that it begins under the soil and so it can be hard to spot it until it's too late. As it spreads, it moves into the lower leaves, and you can possibly catch and stop it then, but a slight discoloration and odd feeling aren't easy to notice if you don't have a good maintenance routine. To begin with, avoid overwatering your crops. Always check the soil with a finger to ensure that it is dry. With most crops, you are better off erring on the side of drought rather than drowning.

But to prevent disease and pests, you need to get in and look at your crops daily. Being with your crops is the best way to get a feeling for how they are supposed to look, smell, and act. Yes, act. If they are falling over when they aren't supposed to, they are acting weird. If you notice holes in the leaves or discoloration around the edges, then there's something wrong. If they begin to smell wrong, this may be a sign of disease, but this particular sense is usually the least effective. I've only been able to catch disease in a crop once this way, but once is enough to earn a mention.

Spend time with your crops, and when you do, you can promote further health. Spray them with a treatment of neem oil every week. This is a natural product that does no harm to humans or pets, but it tastes disgusting to pests and helps to prevent infestation. You should also bring along a tissue and wipe the underside of the leaves as you move through the crop. If the tissue comes back with streaks of blood, then there are pests hanging out, trying to hide from your sight. The best way of dealing with pests when growing outside is to blast the plants with water, though aim it so that it goes away from the soil and not down into it. That will knock away a bunch and reduce the number of pests you need to kill. Next, you can release beneficial bugs like ladybugs. These bugs feed on the pests that feed on your crop. They will eventually head out to find their next meal once they clear away the problem.

As you are checking your crops daily, also remove any dead plant matter that has fallen off. While this can be added to the compost pile for later use, it is important to remove it at this point as it offers a place for disease and pests to grow. Removing it gets rid of the chance that it will invite disease into the field.

Maintaining Your Livestock

As mentioned, the biggest issue with livestock will be health issues. Always look in the eyes of your livestock to see if they are leaking any fluid. Feel the udders of your cows and goats to see if they're overly hot. Pay attention to how much they are eating and see if their poop is the right consistency. With bees, check the trays to see if the queen is still laying eggs, and to ensure that the larvae are surviving and growing properly. If you spot issues with any of these, then you will want to reach out to a vet or consult your local beekeepers' society.

But beyond health issues, there are ways that we can maintain our livestock to reduce the frequency of problems. For one, we can clean out the poop from their homes so that they aren't living in filth. Chickens, cows, goats, and bees are all negatively affected by living in their own dirt. It isn't much fun to have to shovel away animal poop, but it can be used to help keep your fields in healthy shape, so it isn't all bad.

Make sure that you are feeding your animals a healthy diet. If they aren't getting the nutrients they need, they'll get just as sick as your crops do. You should also make sure that they always have clean water. You might purchase a large water trough or container, but you shouldn't just fill it and forget it. It is essential that you check it daily to make sure that there are no feces or other contaminating materials present. Basically, if you wouldn't drink from it, then your livestock shouldn't.

You should also brush and maintain the physical appearance of your livestock. This touch is left out when you move up to larger farms, but it is very useful for raising the healthiest livestock possible. Brush off dirt and grime from them. Not only will this improve their appearance, but it will keep them cleaner and give you a chance to get in close and really observe the way they are acting and how they look. By taking the time to maintain their appearance, you also give them the equivalent of a doctor's check-up. Do this at least once a week, along with checking them daily, so that you can spot issues quickly.

If you spot an issue with your livestock, gather as much information about them as you can. Check their heartbeat, check their eyes, nose, and ears. Check their udders. Watch them eat and see if they show a lack of appetite. See how they interact with the other animals. All of this information will help your vet to diagnose the problem so that you can treat your livestock and get them back up to full health.

Chapter Summary

- While there are vets for your livestock, the only doctor your crops have is you.

- Overwatering your plants promotes the development of root rot, and this will kill them quickly. Only water when the soil is dry. Follow the recommended watering schedule for your particular crop.

- You should be checking your crops daily to get a sense of what is normal for them. That way, you can catch issues quickly because you notice when something stands out.

- Check your crops for signs of infestation.

- Apply neem oil to your crops once a week.

- If you have an issue with pests, then blast them off with a hose and introduce beneficial insects to the garden to feed on the pest.

- Remove dead plant matter from the fields to deny disease and pests a place to breed.

- You should be checking your livestock daily for signs of sickness, such as discharge from the eyes and ears.

- Bees should be checked to see if the queen is still producing eggs.

- Watch your animals as they eat to ensure there are no problems with their diet.

- Keep a close eye on your livestock's water. Contaminated water can spread disease.

- Taking care of your livestock's fur will keep them looking their best, but it also lets you get in and give them a once-over to see if there are any issues health-wise or any surprise injuries you hadn't seen.

FINAL WORDS

So there you have it, everything you need to start your mini-farm. Remember that farming techniques need to be adapted to fit the size of your farm, as following the techniques used for bigger farms is a recipe for losing money. To that end, let's take a brief look at the recommendations we've covered.

In chapter one, we looked at why you would want to start a mini-farm in the first place. If you've read this far, then I'm sure you already plan to start one, and so a recap of this chapter would just be beating a dead horse.

In chapter two, we got our hands dirty by looking at the different approaches we can take to grow our crops. We looked at monocropping and saw that it was a terrible option unless done in raised beds. We also saw that raised beds are generally the best approach we have to mini-farming and so I highly recommended you use them in your mini-farm. If you stick with growing in the ground, you know from our discussion that crop rotation is an absolute must. You also know that you can maximize your space by either using mixed cropping or intercropping. Of course, these assume that you have access to the outdoors, and not everyone does. For those without much outdoor space, we looked at how hydroponics can be used to start a mini-farm indoors. Chapter three looked at profitable vegetables that we can grow, such as potatoes, tomatoes, cabbages, onions, and herbs.

Chapter four saw us move from discussing crops to discussing livestock. Whether it is cattle, chickens, goats, or bees, these animals can make us a lot of money through the various products they produce. To get a sense of how we can profit from them, we spent chapter five looking at the products or services that we can offer. Finally, chapter six saw us look at pest prevention and maintenance with the goal of keeping our mini-farm healthy and productive.

You have all the information you need to start planning and making your mini-farm dream into a reality. But I want to leave you with one last thought before you leave.

Just because we are talking about mini-farms, you shouldn't think that these methods can't apply to larger farms. When starting out, it is best to begin with a mini-farm because it will keep your costs down and allow you to turn a profit much quicker. As your farm grows, you shouldn't feel compelled to stay small. If you are lucky and demonstrate that you care about your farm, your crops, and your livestock, then you can grow your mini-farm into the next major farming business. Just remember to keep to the organic and ethical guidelines that we discussed. It might be more profitable, for example, to stuff chickens together tightly. But, not only would that be unethical, anyone doing it would find a reduced quality in the product and, eventually, they'd lose customers. By staying ethical and organic, you ensure that your farm will continue for years to come.

Now get out there and start growing!

www.ingramcontent.com/pod-product-compliance
Lightning Source LLC
Chambersburg PA
CBHW050313120526
44592CB00014B/1888